Steve Parish
PUBLISHING

A Wild Australia Guide

KANGAROOS
& their relatives

STEVE PARISH & KARIN COX

Contents

Introduction 4
  Identifying Macropods 6
  Macropods in Motion 8
  The Macropod Life Cycle 10

**Potoroos, Bettongs &
the Musky Rat-kangaroo** 12
  Long-nosed Potoroo 14
  Long-footed Potoroo 16
  Gilbert's Potoroo 17
  Burrowing Bettong 18
  Southern Bettong 20
  Northern Bettong 21
  Brush-tailed Bettong 22
  Rufous Bettong 23
  Musky Rat-kangaroo 24

Hare-wallabies 26
  Spectacled Hare-wallaby 28
  Rufous Hare-wallaby 30
  Banded Hare-wallaby 31

Pademelons 32
  Tasmanian Pademelon 34
  Red-legged Pademelon 36
  Red-necked Pademelon 38

The Quokka 40
  Quokka 41

The Swamp Wallaby 42
  Swamp Wallaby 43

Tree-kangaroos 44
  Bennett's Tree-kangaroo 46
  Lumholtz's Tree-kangaroo 47

Rock-wallabies 48
  Short-eared Rock-wallaby 50
  Monjon 52
  Nabarlek 54
  Mareeba Rock-wallaby 56
  Godman's Rock-wallaby 58
  Herbert's Rock-wallaby 59
  Allied Rock-wallaby 60
  Unadorned Rock-wallaby 61
  Sharman's Rock-wallaby 62
  Cape York Rock-wallaby 63
  Black-footed Rock-wallaby 64
  Brush-tailed Rock-wallaby 66
  Yellow-footed Rock-wallaby 68
  Proserpine Rock-wallaby 70
  Purple-necked Rock-wallaby 72
  Rothschild's Rock-wallaby 73

Nailtail Wallabies 74
  Bridled Nailtail Wallaby 76
  Northern Nailtail Wallaby 78

**Wallabies, Wallaroos
& Kangaroos** 80
  Agile Wallaby 82
  Black-striped Wallaby 84
  Tammar Wallaby 86
  Western Brush Wallaby 88
  Parma Wallaby 90
  Whiptail Wallaby 92
  Red-necked Wallaby 94
  Antilopine Wallaroo 96
  Black Wallaroo 98
  Common Wallaroo 100
  Western Grey Kangaroo 102
  Eastern Grey Kangaroo 104
  Red Kangaroo 106

Glossary 108
Index 110
Links & Further Reading 111

Introduction

Few other creatures symbolise Australia quite as successfully as kangaroos and wallabies. These odd-looking macropods, with their bounding gait, long tails, pouch young, boxing stance and hardy disposition, have enthralled generations of Australians, as well as visitors to these shores. Three families of marsupials in the superfamily Macropodoidea are commonly referred to as macropods. Small, omnivorous potoroos and bettongs (in the family Potoroidae), along with the Musky Rat-kangaroo in the family Hypsiprymnodontidae, are also known as macropods, as are those in the larger family Macropodidae, which includes herbivorous kangaroos and wallabies.

Fossil evidence links Australia's macropod species with earlier possum-like animals that lived in trees. Over time, the continent's expanding grassland habitats saw a decline in tree-climbing body types and the rise of terrestrial forms better able to exploit these changing environmental conditions. Gradual changes furnished these animals with body features more suited to open terrain. Various models were road-tested along the evolutionary highway until a highly efficient body plan emerged and became standard equipment for most typical Australian macropods. Features included short forelimbs, powerful hindlimbs and long back feet, giving these mammals an upright stance and a bounding gait. Today, not all macropods favour a life on the ground and in the open. Some species, such as tree-kangaroos, have regressed to a life in the trees and developed compact bodies and arboreal features similar to those of modern possums.

Fifty living macropod species are found on the Australian mainland or on adjacent islands. This book features all the species currently recognised. Prior to Europeans arriving on these shores, there were a further six species of these mammals in existence — unfortunately, these are now extinct. Even today, many of Australia's macropods are listed as vulnerable or endangered — especially the smaller rock-wallabies, hare-wallabies, bettongs and potoroos, which have become easy pickings for introduced predators such as foxes and cats.

Left: Kangaroos and wallabies were crucial sources of food and skins for Aboriginal peoples and are frequently depicted in their artwork, songs and dances.

Macropods' specially adapted body shape has allowed them to enjoy wide distribution over the Australian continent — with different species being spread over most of the States and Territories. Not surprisingly, given their proliferation, the kangaroo, especially, has become an Australian icon. It features on the famed boxing kangaroo flag, lends its image to Australia's national airline, Qantas, and is depicted on tourist memorabilia throughout the nation. Macropods' wide distribution and, for many species, abundance, also sees them represented on road signs throughout Australia, warning motorists of possible encounters with fast-moving macropods, or with those browsing roadside verges where the grass is often lush.

FEATURES OF A MACROPOD

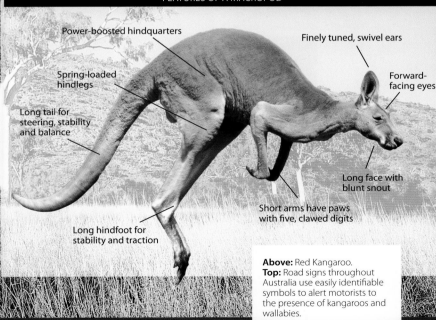

- Power-boosted hindquarters
- Finely tuned, swivel ears
- Spring-loaded hindlegs
- Forward-facing eyes
- Long tail for steering, stability and balance
- Long face with blunt snout
- Short arms have paws with five, clawed digits
- Long hindfoot for stability and traction

Above: Red Kangaroo.
Top: Road signs throughout Australia use easily identifiable symbols to alert motorists to the presence of kangaroos and wallabies.

Identifying Macropods

Correctly identifying macropods in the wild takes practice. A good eye for detail and knowledge of preferred habitats and distribution are essential. The maps in this *Wild Australia Guide* will help you to determine the likelihood of a given species occurring in a particular region. This guide also provides multiple images taken from different angles, as well as text that highlights distinctions between similar species to help you tell them apart. If you are unsure of a species, try using the following criteria to help you identify it in the wild:

- coat colour and distinctive markings;
- whether you are near a particular species' preferred habitat;
- the macropod's stance and hopping posture;
- the macropod's size;
- the shape and size of body parts;
- whether a macropod is behaving in a way common to that particular species.

Top, left to right: Although there are overlaps in range, rock-wallabies are usually identified by their distribution — Black-footed Rock-wallaby; Yellow-footed Rock-wallaby; Allied Rock-wallaby; Mareeba Rock-wallaby.

Centre, left to right: Different wallaby species have distinctive skull shapes and head markings — Western Brush Wallaby; Red-necked Wallaby; Swamp Wallaby; Whiptail Wallaby.

Bottom, left to right: Variations in fur colour and overlapping ranges can make identification of larger macropods tricky — Eastern Grey Kangaroo; Western Grey Kangaroo; Red Kangaroo; Antilopine Wallaroo.

IDENTIFICATION ISSUES

Relying on coat colour and pattern alone can pose problems. Light diffusion at different times of day can easily confuse the human eye, as demonstrated by the pictures above. The four images of the Common Wallaroo (*Macropus robustus*) shown above left (top to bottom), differ due to light, age and gender, and could easily lead to confusion with other large macropods. One of this species' most distinguishing features is its rounded ears. Four different photographs shown above right (top to bottom) of the Whiptail Wallaby (*Macropus parryi*) appear to be different species under differing lighting conditions. Each species may also have genetic colour variations or regional differences; age, gender and climate can also affect fur colour and density.

Macropods in Motion

Hindlimbs that cannot move independently of each other prohibit most kangaroos and wallabies from walking or running on all fours (exceptions are tree-kangaroos and the Musky Rat-kangaroo). Strangely, this inability is purely a quirk of the macropod nervous system — when swimming, kangaroos are able to kick their legs alternately. To travel slowly on land most macropods use all limbs, including their tails. To move at a faster pace, they prefer an efficient two-footed hop.

Astonishingly, the macropod's awkward body shape gifts it with a swiftness lacking in all other Australian marsupials. Muscular back legs (combined with a sturdy, balancing tail) propel the kangaroo or wallaby forward at a remarkable pace. The speed record-holder to date is an Eastern Grey Kangaroo clocked at 64 km/h, although even greater speeds may be possible. Heights, too, are easily scaled. A male Red Kangaroo has been observed easily clearing a 3.1 m high stack of timber. Hopping, although a seemingly exhaustive method of locomotion to humans, is an extremely efficient, energy-saving way to travel. At speeds greater than 15 km/h, hopping minimises oxygen use, using considerably less oxygen per bound than running or galloping. With each leap, the powerful leg muscles and flexible tendons store and release energy in much the same way as an elastic band. While hopping, a macropod also decreases the effort required to breathe, as air is forced in and out of its lungs as its gut flops up and down.

PENTAPEDAL WALK

To move slowly in a pentapedal (or five-limbed walk), a macropod supports its body on its front limbs and balancing tail while swinging the hindlimbs forwards with a pendulum-like motion.

BIPEDAL HOP

For swifter movement, a macropod uses its power-packed, muscular hindlegs to hop. The long tail provides balance, while the arms and head are usually positioned in a way that best streamlines the body. The Western Brush Wallaby (above) uses its particularly long tail to balance its powerful, fully arched gait.

POSTURE & GAIT

Although it may seem that a fleeting glimpse of a fleeing macropod leaves little chance for accurate identification, observing gait and posture can often aid identification. The length of the macropod's stride, the frequency and height of its hops and the position of the head and arms provide solid clues to identifying the species. The long legs of the Whiptail Wallaby (above, left) and the large Red Kangaroo (second from left) are set well back on the body, giving the forebody a characteristic arch. The shorter-legged Agile Wallaby (second from right) and Eastern Grey Kangaroo (right) both employ a shorter stride. Both also adopt an upright hopping posture, but the position of the forelimbs is distinctly different.

The Macropod Life Cycle

The life cycle of macropods differs from species to species, although most species enjoy remarkable fertility, with many macropod females able to come into oestrus throughout the year. To ensure the survival of their young, females of many kangaroo and wallaby species benefit from their ability to "pause" reproduction — giving birth only when environmental conditions favour the joey's survival. In times of drought, females do not come into oestrus; instead, they control the rate at which an embryo develops by an incredible phenomenon known as "embryonic diapause". This enables the female to suspend development of the embryo, now called a "blastocyst", in her uterus (sometimes for up to 33 weeks) until the time is right for its further development and birth. During this time, the mother may also be rearing pouch young. Development of the dormant embryo will resume if her joey dies or grows old enough to survive outside the pouch. Embryonic diapause is a particularly useful strategy in harsh environments where infant mortality is high. When food and water are plentiful, the gestation period is usually between 29 and 38 days (depending on the species), and females frequently mate again soon after giving birth. During a good season, a macropod mother may have an unweaned joey at heel, a suckling joey in the pouch and an embryonic joey in waiting.

Above: A Tammar Wallaby embryo tears through the foetal membrane (top left), and then begins its climb to the pouch (top right and above left), where it begins to feed (above right). Newborn joeys are tiny, furless and weigh less than 2 g.
Opposite: Bridled Nailtail Wallabies mating.

Soon after a doe gives birth, the underdeveloped, jelly-bean-sized joey begins an amazing climb to survival. Pulling itself through the dense fur that surrounds its mother's cloaca, it makes its way to her pouch (or marsupium), where it attaches to one of four teats and begins to suckle. This instinctive journey takes around three minutes for a newborn Red Kangaroo. The joey will then spend six months in the pouch, developing fur and growing stronger, until it begins to make its first tentative journeys out of its safe, warm abode. From 6–9 months of age, the joey may make regular forays out of the pouch and will eventually receive its eviction notice just before another tiny joey embryo is born.

Potoroos, Bettongs & the Musky Rat-kangaroo

Small, omnivorous rat-kangaroos, potoroos and bettongs belong to the family Potoroidae — the members of which retain some of the features of their long-extinct mammalian ancestors (such as weakly prehensile tails).

Unlike all other macropod species, members of the family Potoroidae build nests (the materials for which they carry bundled in their tails). They favour habitats that provide a thick, protective understorey as well as abundant food and shelter. Plants, insects and fungi comprise most of their diet. To cope with their omnivorous menu, potoroid dentition includes upper canines and fixed molars, which are not found in the larger herbivorous macropods.

Despite their differences, potoroos, bettongs and the Musky Rat-kangaroo still exhibit the typical macropod's long hindfoot with fused grooming toes; hence they are commonly accepted as macropods, although the more distantly related Musky Rat-kangaroo, which lacks muscular hindlimbs, gallops with a quadrupedal gait rather than hops like most other macropods.

This family of macropods has particularly suffered as a result of human habitation.

Top: The now-extinct Desert Rat-kangaroo (*Caloprymnus campestris*) immortalised in artwork by John Gould.
Right: Another casualty of European settlement was the Broad-faced Potoroo (*Potorous platyops*). **Opposite:** Long-nosed Potoroo (*Potorous tridactylus*).

Long-nosed Potoroo *Potorous tridactylus*

*The Long-nosed Potoroo's erroneous species name (*tridactylus*) refers to the initial belief that this species had only three toes. In fact, the second and third toes are conjoined — giving the Long-nosed Potoroo four toes like other members of its kin. This species was one of Australia's first recorded small mammals, with an illustration and description given in 1789 by Governor Arthur Phillip at Botany Bay.*

FEATURES: Slightly smaller than the Long-footed Potoroo, with a furrier tail and shorter hindfoot. Sometimes a white tip at the end of the tail. Considerable variations in colour (rufous brown to greyish with paler fur on the underside) and size are found over their distribution, with larger, longer nosed specimens found in Tasmania. On the whole, the eponymous snout is rather narrow, with a bare patch above the nostrils.

DIET & HABITAT: This species is rather widely distributed, especially across rainforest, coastal heath and eucalypt forest through south-eastern Australia from Queensland to eastern Victoria and in Tasmania and some Bass Strait Islands. They are slightly less dependent on a fungal diet than the Long-footed Potoroo and will also eat roots, tubers and invertebrates.

BEHAVIOUR: Mostly solitary and nocturnal, although may sometimes congregate in small groups. This species sometimes digs shallow holes in the ground to search for roots, but cannot be said to be a true burrower.

BREEDING: Can breed throughout the year but most births (after a gestation period of 38 days) occur from late winter to early spring. A single offspring is produced twice a year and remains in the pouch for approximately four months. Female sexual maturity is reached at 8–10 months, with male sexual maturity taking slightly longer.

THREATS: Dingoes, dogs, cats, owls, foxes and habitat clearing.

Above: Fungi makes up a larger part of this species' diet and such food requires a specialised digestive system.

DIET: Fungi, roots and tubers, insects and soft-bodied invertebrates

LENGTH: HB 34–38 cm; T 19.8–26.2 cm
WEIGHT: 660–1640 g
STATUS: Vulnerable in parts of range

Above: Long-nosed Potoroos in Tasmania have longer, narrower snouts than those living in Queensland. Tasmanian specimens are also more likely to have white-tipped tails.

Long-footed Potoroo *Potorous longipes*

First recorded in Victoria's Gippsland area in the late 1960s, the endangered Long-footed Potoroo was only formally described in 1980. It now has fragmented populations in East Gippsland, the north-eastern Victorian Alps (between Mt Feathertop and Mt Buller) and in Bondi State Forest in south-eastern New South Wales. A small captive population at Healesville Sanctuary in Victoria is helping to provide ongoing research into the behaviour and reproduction of this species.

FEATURES: The long hindfoot is this species' main distinguishing feature. Fur is grey-brown on the back, with a paler grey underneath the body. The long, thick tail is sparsely furred.

DIET & HABITAT: Temperate rainforest, wet sclerophyll forest and riparian forest with heavy undergrowth of ferns, shrubs and sedges, and plenty of underground fungi — a major component of this species' diet — is the preferred habitat of the Long-footed Potoroo. Part of its range also takes in altitudes of 150–1000+ m. Long-footed Potoroos are essentially fungivores, but it is thought they may eat small invertebrates and plant matter as well.

BEHAVIOUR: This species is nocturnal, sheltering during the heat of the day in a loosely constructed nest of ferns or grasses. Locomotion consists of short bipedal hops. When distressed, this species may vocalise with a quiet "kiss kiss" sound, which is also a common maternal vocalisation.

BREEDING: Sexual maturity is reached at around two years of age, after which females can have two to three young per year. Gestation period is believed to be about 38 days, after which a single offspring is born and nurtured in the pouch for around 20–22 weeks.

THREATS: Dingoes, foxes, feral dogs and cats, and habitat clearing (in many parts of its range outside national parks and reserves) are major threats.

Above: Because the bulk of its diet consists of fungi, Long-footed Potoroos play an important ecological role in the dispersal of spores.

DIET: Largely reliant on underground and sub-underground fungi

LENGTH: HB 38–41.5 cm; T 31.5–32.5 cm
WEIGHT: 1.6–2.2 kg
STATUS: Endangered

Gilbert's Potoroo *Potorous gilbertii*

With a total of about 40 individuals residing in a 1000 ha pocket in Two Peoples Bay Nature Reserve, Western Australia, Gilbert's Potoroo is Australia's most critically endangered mammalian species. For more than 100 years, this small potoroo was believed to be extinct, until a chance finding in 1994 clarified history. Since 2005, successful relocation has resulted in a second colony of about thirteen individuals on Bald Island off the Western Australian coast.

FEATURES: Somewhat bandicoot-like, with grey to brown fur and a lightly furred tail. Small rounded ears. The slender snout curves down at the tip.

DIET & HABITAT: Gilbert's Potoroo prefers *Melaleuca* shrubland (particularly habitat dominated by *M. striata*) where a dense canopy shelters thick sedges and ground cover through which numerous protective "runways" are constructed. Gilbert's Potoroo is heavily reliant on the fruiting bodies of underground fungi — more than 90% of its diet is comprised of "truffles", with berries, insects and some fleshy seeds eaten only occasionally.

BEHAVIOUR: When moving slowly, Gilbert's Potoroos use their forepaws for balance, but hop using the hindlimbs only. Each animal occupies a home range, with the males' home ranges (usually around 15–25 ha) overlapping those of females. They are nocturnal, resting by day in well-hidden, shallow depressions underneath sedges.

BREEDING: The estimated age of sexual maturity is about a year for females, with males maturing later at around two years of age. Females produce a single offspring at any time of year, following a gestation period of 4–6 weeks. Females can employ embryonic diapause if necessary. Young leave the pouch at about four months and are independent at six months of age.

THREATS: Burning off of shrubland probably factored in their decline. Dingoes, foxes, feral dogs and cats.

Above: Considered extinct for more than 100 years, Gilbert's Potoroo was rediscovered in Two Peoples Bay Nature Reserve (WA) in 1994.

DIET: Heavily reliant on the fruiting bodies of underground fungi

LENGTH: HB 27–29 cm; T 21.5–23 cm
WEIGHT: 900–1100 g
STATUS: Critically Endangered

Burrowing Bettong *Bettongia lesueur*

This fascinating mammal has the honour of being the only macropod species to construct complex burrows, or warrens, where up to 100 individuals enjoy a sociable lifestyle. Unfortunately, this species has been driven to local extinction on the mainland of Australia, and now exists only on Bernier, Dorre, Barrow and Boodie Islands off the central coast of Western Australia.

FEATURES: Dense greyish-brown fur has a slight yellowish tinge on the back, becoming pale grey on the body's underside. The head is small and the nose blunt. Small ears are short and rounded. Tail is thick and sparsely furred, sometimes with a white tip.

DIET & HABITAT: Once widespread, it prefers hummock grasslands and scrub in semi-arid and arid areas where condensed, loamy soils allow for the construction of elaborate burrows. Their partly subterranean lifestyle suits their diet of roots, tubers and underground fungi. They also eat fruit, seeds, flowers, arthropods and some carrion.

BEHAVIOUR: Burrowing Bettongs are a gregarious and vocal species, emitting a number of communicative squeaks, squeals, grunts and hisses. They are nocturnal, resting by day in their burrow systems. Locomotion is always bipedal — a feature that readily distinguishes them from rabbits, which were recorded sharing warren systems with Burrowing Bettongs in former mainland populations.

BREEDING: Breeding occurs throughout the year, with a single joey produced three times a year after a gestation of 21 days. Maturity is reached at a relatively early age (around seven months).

THREATS: Foxes and feral cats have had a devastating effect on this species, which now inhabits less than 0.01% of its former range. Conservation programs are reintroducing this species to the mainland at Arid Recovery Reserve (SA), Heirisson Prong (WA), and Scotia Sanctuary (NSW).

Above: The Burrowing Bettong is a social macropod species. Dozens of bettongs may share the same system of burrows.

DIET: Fungi, tubers, roots, small invertebrates, flowers, seeds and carrion (in some cases)

LENGTH: HB 28–36 cm; T 21.5–30 cm
WEIGHT: 900–1500 g
STATUS: Vulnerable in parts of range

Above: The Burrowing Bettong enjoys a varied diet. Besides foraging for roots, fungi, seeds and invertebrates, this species has been observed eating carrion on the beach.

Southern Bettong *Bettongia gaimardi*

Also sometimes referred to as the Tasmanian or Eastern Bettong, this species probably became locally extinct on the Australian mainland in the early 20th century. It now exists only in Tasmania, where it enjoys relative security compared with many other bettong species.

FEATURES: The brownish-grey upper body is occasionally lightly streaked with buff or white around the flanks, face and shoulders. Underneath, the body is cream to greyish-white. The furry tail is a reddish-grey, tapering to black at the end — sometimes with a white tip. The ears are small, lightly furred and well rounded.

DIET & HABITAT: Southern Bettongs inhabit dry, grassy woodlands and open sclerophyll forests in much of south-eastern Tasmania, including Bruny and Maria Islands.

BEHAVIOUR: Daylight hours for this strictly nocturnal species are spent in a tightly constructed nest of bark and grass (approximately 30 cm long by 20 cm wide). Nests are usually well concealed in slight depressions dug beneath the shelter of a fallen limb or among short shrubby bushes and grass tussocks. Each individual has a home range of 65–135 ha and may travel some distance from the nest to suitable feeding areas. Southern Bettongs are considered solitary, although on occasion two individuals may share the same nest. Captive males are known to be rather aggressive with other males.

BREEDING: Sexual maturity is reached at around one year of age. Births are recorded year-round with a single joey born after a gestation period of 21 days. The joey stays in the pouch for approximately fifteen weeks and is then weaned 40–60 days after leaving the pouch. Two or three offspring may be produced per female per year.

THREATS: The absence of the Red Fox from Tasmania (up until recently) has probably been this species' saving grace. Rabbits, too, are not well established in Tasmania, thus preserving the grassy woodland favoured by the Southern Bettong. Habitat clearing and feral cats are the two main threats faced.

Above: Now extinct on mainland Australia, the Southern Bettong survives in reasonable numbers in Tasmania.

DIET: Mainly underground fungi, roots and bulbs and some seeds

LENGTH: HB 31.5–33.2 cm; T 28.8–34.5 cm
WEIGHT: 1.2–2.25 kg
STATUS: Near Threatened (IUCN Red List)

Northern Bettong *Bettongia tropica*

Some taxonomists originally classified the endangered Northern Bettong, which occupies only a small range in north Queensland, as a subspecies of the more common Brush-tailed Bettong. However similar, the two are currently believed to be separate species and the vast distance in distribution between the two would appear to justify this classification.

FEATURES: Similar in appearance to the Brush-tailed Bettong (although unlikely to be confused due to their different distributions) with a grey body and lighter beige underside. The tail is somewhat greyer than that of the Brush-tailed Bettong but retains the crest of dark hair at the tip.

DIET & HABITAT: Populations are currently restricted to just three areas of tall and medium sclerophyll forest and nearby rainforest (mostly at altitudes of 450 m or more) in north-east Queensland — Mt Carbine Tableland, Lamb Range and Coane Range. Like most other bettong species, their diet consists mainly of fungi (making up approximately 43% of the diet), tubers and grass, especially Cockatoo Grass (*Alloteropsis semialata*), although seeds and invertebrates are consumed in small quantities.

BEHAVIOUR: Strictly nocturnal and largely solitary, these bettongs rest during the day in nests of woven grass matter or beneath grass-trees. Each individual constructs as many as three nests and uses them sporadically when in the area.

BREEDING: Females are receptive throughout the year. A single offspring is produced at each birthing and when conditions are favourable up to three joeys a year are born. Other aspects of their reproductive behaviour, such as gestation time, are similar to those of the Brush-tailed Bettong.

THREATS: Fire appears to affect the subsequent availability of underground fruiting bodies of fungi — a major element of this species' diet. Habitat change and deforestation, as well as predation by foxes and feral cats could vitally impact on this limited and endangered mammal.

Above: Northern Bettongs are a solitary species. Individuals use up to three different nests throughout their home range.

DIET: Mostly fungi, tubers, roots, grass and limited seeds and insects

LENGTH: HB 26.7–40.4 cm; T 31.7–36.4 cm
WEIGHT: 900–1400 g
STATUS: Endangered

Brush-tailed Bettong *Bettongia penicillata*

The Brush-tailed Bettong — or Woylie as it is also commonly known — is easily distinguished by the switch-like "brush" of dark hair on its tail. Like most other bettong species, it builds nests — gathering the material for its home in its weakly prehensile tail.

FEATURES: Fur is greyish-brown with some silver speckling, becoming a yellowish to beige colour on the underside of the face and around the thigh and flanks. Underneath, the body is cream. The ears are small and rounded and the highly distinctive tail is reddish, terminating in a crest of dark, bristle-like hair.

DIET & HABITAT: Brush-tailed Bettongs prefer open dry sclerophyll forest, preferably with a thick understorey where they can weave a nest of grass, bark and twigs. Although once widely distributed in arid habitats across south and north-western Australia, they are now common only around Dryandra, Perup and Tutanning in South-West Western Australia. Like most bettongs, they dine primarily on fungi, bulbs, tubers, seeds and insects.

BEHAVIOUR: During the day, this nocturnal marsupial shelters in a domed nest. Although solitary, Brush-tailed Bettongs occupy home ranges that may overlap. A particularly distinctive aspect of this species' behaviour is its unusual zig-zagging hop, made with the head held low, tail extended and body rigidly arched.

BREEDING: From around 24 weeks of age, females come into oestrus every 100 days and give birth to a single joey at a time (occasionally two, although only one survives). The joey stays in the pouch for just over twelve weeks, after which time it will continue to share its mother's nest until another joey usurps its place at its mother's side. Two to three offspring are born each year.

THREATS: Foxes and habitat destruction (land clearing) have limited its range to Western Australia. Successful reintroduction programs continue in South Australia.

Top to bottom: As its common name suggests, this species' bristly tail is distinctive; Small rounded ears are also a feature of this species.

DIET: Fungi, tubers, roots, small arthropods

LENGTH: HB 30–38 cm; T 29–36 cm
WEIGHT: 1.1–1.6 kg
STATUS: Conservation Dependent (IUCN Red List)

Rufous Bettong *Aepyprymnus rufescens*

Although the Rufous Bettong is probably Australia's most widely distributed potoroid species, it has still suffered a decline in numbers since European settlement. This species is another of the nest-builders, excavating shallow depressions and constructing dome-shaped nests of fibrous plant material over the top.

FEATURES: Shaggy, greyish-brown fur with a distinctive reddish (rufous) tinge and longer, protruding silver hairs merge to a pale silvery-grey on the underside of the body. The ears are triangular and longer than those of other living bettong species. Furless skin on the eyes and a lightly furred strip from nostril to eye are characteristic of this species (and the Southern Bettong).

DIET & HABITAT: Rufous Bettongs live in various habitats — from wet sclerophyll forest to coastal woodland and dry, semi-arid woodland. Rufous Bettongs usually emerge from the nest about a half hour after sunset to forage for roots, tubers and flowers. They will also browse on grasses and herbs, and have been known to eat carrion. This diet provides sufficient water, meaning that they only need to drink during times of drought.

BEHAVIOUR: Each individual may create around five nests in its 45–110 ha home range, occupying the nests temporarily when in the area. They are nocturnal and were originally thought to be solitary, although it now appears they may form loose social groups of one male with a few associated females. When alarmed, they issue a hiss or loud chainsaw-like alarm sound.

BREEDING: Females come into oestrus every three weeks from around eleven months of age. A single joey is born after a 22–24 day gestation period and leaves the pouch at sixteen weeks.

THREATS: Dingoes, foxes, feral cats and competition with rabbits for food could further impact on the Rufous Bettong's conservation status.

Above: Reddish fur with silver emergent hairs is a distinguishing feature of the Rufous Bettong. The ears are typically pointed.

DIET: Omnivorous, grazing on grass and herbs, roots, tubers, flowers, carrion

LENGTH: HB 37.5–39 cm; T 33.8–38.7 cm
WEIGHT: 3–3.5 kg
STATUS: Vulnerable (NSW); Threatened (Vic)

Musky Rat-kangaroo *Hypsiprymnodon moschatus*

Australia's smallest macropod is also one of the most fascinating. This species bears a number of distinguishing features and is only loosely related to other potoroid species, being the sole living member of the subfamily Hypsiprymnodontidae. Major differences include having five toes (with the first being opposable), occupying a rainforest habitat, and being the only macropod to give birth to two (sometimes three) offspring at once.

FEATURES: This species is much smaller and darker coloured than other potoroids. Chocolate brown, slightly rufous-tinged fur gives way to a greyer head. A whitish band extends from the underside of the throat to the abdomen. The five-toed feet are black and the tail slim, furless (almost scaly in appearance) and relatively short.

DIET & HABITAT: Rainforest and montane rainforest in north-eastern Queensland is this species' preferred habitat. Here it forages in the leaf litter for rainforest fruit (such as palm berries, figs and quandongs), as well as fungi and insects.

BEHAVIOUR: The species takes its common name from the pungent, musky scent that accompanies it. Unlike most other potoroids, it is diurnal or crepuscular (active at dusk and dawn) and may congregate in small groups. At night, it shelters in a loosely constructed nest. While this species is largely terrestrial, the opposable "big toe" on each hindfoot enables the Musky Rat-kangaroo to swiftly leap over and climb through fallen trees. Another unusual feature is this species' bounding quadrupedal gait, further distinguishing it from its macropod relatives.

BREEDING: Sexual maturity is reached at around two years in wild specimens. Giving birth to two offspring in a litter is usual, with most births occurring from October to April. The joeys remain in the pouch for around 21 weeks, after which time they remain in the nest and continue to suckle until weaned.

THREATS: Clearing of land for agriculture and fragmentation of rainforest are a concern for this species.

Above: This unique macropod is unlikely to be misidentified. A distinguishing feature is the opposable first toe of each hindfoot.

DIET: Largely fruit, along with fungi and invertebrates in leaf litter

LENGTH: HB 15.3–27.3 cm; T 12.3–15.9 cm
WEIGHT: 360–680 g
STATUS: Secure in a limited habitat

Above, top to bottom: The Musky Rat-kangaroo is Australia's smallest macropod species; Musky Rat-kangaroo rainforest habitat.

Hare-wallabies

Four species in the genus *Lagorchestes* (in the subfamily Macropodinae), along with one species in the genus *Lagostrophus* (the sole member of the much more ancient subfamily Sthenurinae), are commonly referred to as hare-wallabies. Hare-wallabies are compact, water-efficient macropods with a temperature-control system specifically designed for dry outback conditions. Shallow, scraped-out hollows and tunnels under (or through) tussock or hummock grass and shrubs offer protective cover in arid and semi-arid habitats. To escape the sweltering summer heat, hare-wallabies also sometimes dig shallow burrows up to 70 cm deep.

All species are nocturnal, leaving their cool refuges at night to browse on leaves, grass tips and fleshy, succulent plants. Hare-wallabies do not need to drink because sufficient water is provided by their diet. They produce little urine and do not utilise their evaporative-cooling system until the air temperature exceeds 30 °C. Despite their suitability to harsh conditions, the European settlement of Australia and subsequent introduction of feral species have severely impacted on hare-wallaby numbers. Two of the four *Lagorchestes* species have succumbed to extinction, while the remaining species have all declined in number.

Above: John Gould's illustration of the extinct Eastern Hare-wallaby.
Opposite: Once common across Central Australia, the Rufous Hare-wallaby is an important species in Aboriginal mythology.

Spectacled Hare-wallaby *Lagorchestes conspicillatus*

Considered the only hare-wallaby that remains secure, the Spectacled Hare-wallaby is probably also the most easily identifiable hare-wallaby species thanks to the rufous rings around its eyes (giving the species its "spectacled" appearance). Its distribution takes in parts of northern Queensland, the Northern Territory and Western Australia (including Barrow Island).

FEATURES: This stocky, thick-set greyish-brown macropod has very obvious rings of reddish-orange fur around its eyes and a red tinge to the fur on its flanks and ears. A white "moustache" is present below the nose and a white stripe extends from the paler buff/white underbelly to the hips. The tail colour is grey-brown, growing darker at the tip.

DIET & HABITAT: Inhabits a number of environments — from mainland tropical hummock grassland to open forest and woodland and the Great Sandy Desert (although its numbers have drastically declined in such arid regions). This species feeds on shrubs, grasses and leaf tips and does not need to drink.

BEHAVIOUR: Spectacled Hare-wallabies are nocturnal and spend the day in tunnels below the grassland to avoid the oppressive heat. When relaxed they move with a slow pentapedal walk, but if alarmed will adopt a bouncy bipedal hop. Up to three of these usually solitary animals may be seen feeding together. Danger is communicated between individuals by a low warning hiss. The only other common vocalisation is a gentle clicking sound issued by the male if a female in oestrus is nearby. A similar vocalisation is also made by mothers to their offspring.

BREEDING: Both males and females are sexually mature at around one year of age. Births of a single offspring occur throughout the year (after a gestation period of 29–31 days), but peak in March and September. During unfavourable conditions, embryonic diapause is practised.

THREATS: Competes with grazing domestic stock in much of its range. Feral cats have contributed greatly to its decline. During drought, lack of food could lead to starvation.

Left and opposite: This species is easily identified by it rufous "spectacles".

DIET: Green tips of grasses, shrubs and herbs

LENGTH: HB 40–47 cm; T 37–49 cm
WEIGHT: 1.6–4.5 kg
STATUS: Near Threatened (IUCN Red List)

Rufous Hare-wallaby *Lagorchestes hirsutus*

Despite being common across the central and western deserts until the 1930s, the Rufous Hare-wallaby's wild range is now restricted to Bernier and Dorre Islands off the Western Australian coast. This species is commonly referred to as the "Mala" and plays a significant role in the Dreaming stories of the Anangu people of Uluru–Kata Tjuta National Park, Northern Territory, where a small captive population of Rufous Hare-wallabies has been reintroduced to a "feral-proof" fenced enclosure.

FEATURES: This species is more petite and dainty than the Spectacled Hare-wallaby and its fur is more consistently light red-orange in colour. (There are also no "spectacles" to speak of.) Fur fringing the ears, top of the head and spine is greyish in colour and the underside of the body and forearms are pale yellowish.

DIET & HABITAT: Prefers spinifex grassland and coastal shrubland where fresh shooting grass, shrubs, herbs and seeding plants are plentiful, especially following seasonal fire. Although this species can survive on a diet of spinifex, less fibrous flora with a higher water content is preferable. In dry times, insects may also be eaten.

BEHAVIOUR: The female Rufous Hare-wallaby grows slightly larger than the male, which is unusual for a macropod. Both males and females are solitary and nocturnal, sheltering by day in hollowed-out scrapes or shallow burrows dug under spinifex tussocks.

BREEDING: Females are fertile from around five months, males from fourteen months. Births occur throughout the year; a single offspring is produced and remains in the pouch for approximately eighteen weeks.

THREATS: Feral cats and human habitation have drastically reduced numbers, causing local extinction in some parts of their former range. Efforts are being made to reintroduce captive-bred groups to the Tanami Desert.

Above: The Rufous Hare-wallaby's red-orange fur camouflages well with aridland soil. Sadly, it is now extinct in most of its former range.

DIET: Green shoots and tips of grasses, shrubs and herbs; sometimes insects

LENGTH: HB 31–39 cm; T 24.5–30.5 cm
WEIGHT: 780–1960 g
STATUS: Vulnerable (IUCN Red List)

Banded Hare-wallaby *Lagostrophus fasciatus*

In actuality only distantly related to the "true" hare-wallabies of the genus Lagorchestes, *the Banded Hare-wallaby occupies a different genus and is believed to be the sole surviving member of the ancient sthenurine kangaroos. Its incisors are unlike those of any other macropod's and its heavily banded rump easily distinguishes from all other hare-wallaby species.*

FEATURES: This species' most obvious distinguishing feature is the darker coloured fur with prominent dark transverse stripes running across its rump. Grey fur on the body has protruding, long, silvery hairs and is dense and somewhat shaggy, giving this species a grizzled appearance. The flanks display a slight reddish tinge and the grey tail is thinly furred with a black crest at the end.

DIET & HABITAT: Although once widespread throughout the wheatbelt regions of Western Australia, this species is now confined to Bernier and Dorre Islands in Western Australia where it is usually found in or around dense stands of *Acacia ligulata* scrub. This species is nocturnal, grazing on grass, legumes and dicotyledonous plants in open areas among scattered pockets of cover.

BEHAVIOUR: Banded Hare-wallabies are solitary — individuals occupy their own home range or territory, with a male's territory overlapping the territories of a number of females. Males are particularly aggressive if they come into contact.

BREEDING: Most births occur towards the end of the Australian summer (in January or February), but some births occur slightly later (in March or April). In tough times, embryonic diapause is used as a reproductive strategy. Young are weaned at around nine months of age and may become sexually mature at twelve months. Usually only one offspring is born per female per year.

THREATS: Feral cats and habitat destruction, coupled with this species' slow reproduction rate, are detrimental.

Above: The Banded Hare-wallaby was one of the earliest macropod species to be recorded. French naturalist and artist Charles Leseuer first depicted it in 1801.

DIET: Grazes on shrubs, spinifex grasses, legumes and dicotyledonous plants

LENGTH: HB 40–50 cm; T 35–40 cm
WEIGHT: 1.5–3 kg
STATUS: Vulnerable

Pademelons

Shy, retiring pademelons are the fringe-dwellers of Australia's eastern forests. Being so reclusive, they usually prefer to make only short-range forays out of the forest, emerging in the late afternoon or early morning to feed on nearby grassy clearings.

Pademelons' low-slung, stout bodies are perfect for propelling them through the thick undergrowth of wet forests. These macropods move quickly with short, bouncy hops, holding the arms close to the body and extending the tail behind for balance. A network of runways aids them in their journeys to and from feeding grounds and Pademelons may travel several kilometres in a night searching for food.

Although pademelons are naturally cautious, a quiet and patient observer may be granted an audience with these shy macropods around campsites, where pademelons often become accustomed to human company. If frightened, an individual will first thump its hindfeet to warn its companions, before springing to safety itself.

Above: A common protective tactic of pademelons is for a pair to take opposing sides when browsing, thus keeping a full field of view for predators.
Opposite: Pademelons could be called the "pocket rockets" of the macropod world. They are small, but can move with great speed when alarmed.

Tasmanian Pademelon *Thylogale billardierii*

Humans must take responsibility for the disappearance of this small, fluffy macropod from the Australian mainland. A victim of the fur and leather trade, it now exists only in Tasmania and the larger islands of Bass Strait.

FEATURES: Thick, fluffy dark brown to grey-brown hair takes on a rufous or buff tint underneath the body. Fur inside the ears is also distinctively red-orange coloured. The face is short and blunt with greyish fur on the cheeks. The tail is also rather squat and thick. Males grow much larger than females.

DIET & HABITAT: Thickly vegetated rainforest, tea-tree scrub and wet sclerophyll forest — along with some areas of open dry sclerophyll forest with a grassy understorey — make up the favoured habitat for this species. In these places they forage for fresh green pick of grasses, herbs and new-growing leaves. In snowy country, the stout forelimbs are used to uncover vegetation buried beneath the snow.

BEHAVIOUR: A nocturnal existence suits this small macropod and it is not often seen by day other than in the early morning and late afternoon. Although solitary, small groups may form in grazing areas, but individuals remain close to the forest edge. It is a shy, silent forest resident and rarely vocalises. Bucks may give the odd guttural hiss of aggression or make clucking sounds when courting a doe in oestrus.

BREEDING: Breeding occurs all year with a peak of births in April, May and June. A single joey is born after a gestation period of 30 days and stays in the pouch for around 28 weeks.

THREATS: Humans exploited this species for fur, leather and meat. It has few predators in Tasmania.

Above: Tasmanian Pademelons are small wallabies that prefer to live in wet forest regions where juicy grasses and shrubs are plentiful. **Opposite, top:** This species spends the day hidden beneath vegetation where dappled light makes it hard to spot. **Opposite, bottom:** A young pademelon enjoys the morning sun.

DIET: Grasses, herbs and leaves

LENGTH: HB 56–63 cm; T 32–48.3 cm
WEIGHT: 2.4–12 kg
STATUS: Secure (Tas)

Red-legged Pademelon *Thylogale stigmatica*

More widespread than its southern cousin, the Red-necked Pademelon, this species is widely distributed along the east coast from northern New South Wales to islands off the tip of Cape York. It is also found in New Guinea.

FEATURES: Short, bristly fur is darker and more rufous all over than that of the Red-necked Pademelon, becoming even redder on the legs (as its common name suggests). The cheeks have a marked white stripe and the underside of the body is a greyish-white. A grey to blackish midline runs between the large, oblong ears and down the forehead. When hopping, the short, thick tail sticks out straight behind.

DIET & HABITAT: Red-legged Pademelons inhabit rainforest and wet sclerophyll forest. They consume a variety of plant matter, from leaves and rainforest fruit to ferns, grasses and agricultural crops (where available). Diet differs depending on habitat — southern specimens consume more fruit and rainforest plants, particularly fig species, while northern individuals tend to move out of the forest to graze on grassy fringe country.

BEHAVIOUR: Red-legged Pademelons are one of the few macropods that can be said to be diurnal. Although active for the best part of 24 hours, they will rest during late afternoon and again around midnight. When resting, they adopt a slumped posture against a tree branch, with the tail pulled forwards between the hindlegs. Red-legged Pademelons are a highly vocal macropod species, making a number of communicative sounds. Maternal "clucking" sounds are imitated by males seeking to mate — uninterested females respond with a loud, hostile rasping noise.

BREEDING: Females reach sexual maturity at approximately 48 weeks of age and then come into oestrus every 29–32 days. Following a gestation period of 28–32 days, a single joey is born and remains in the pouch for 26–28 weeks.

THREATS: Predation by Dingoes, Spotted-tailed Quolls and Scrub Pythons, as well as habitat destruction.

Above: The joey begins to make excursions out of the pouch from 22 weeks and usually exits the pouch entirely from 28 weeks. Around 66 days later, it will be fully weaned.

DIET: Rainforest fruit such as figs, Burdekin Plum, seeds of the Pink Ash, grasses, herbs and ferns

LENGTH: HB 38.6–53.6 cm; T 30.1–47.3 cm
WEIGHT: 2.5–6.8 kg
STATUS: Vulnerable (NSW); Secure elsewhere

Above: Red-legged Pademelons using the forest edge (those in the northern range) are generally paler than those of the rainforest.

Red-necked Pademelon *Thylogale thetis*

The Red-necked Pademelon is a frequent but tentative visitor to forest edges from the south-eastern corner of Queensland to the central coast of New South Wales. It is best seen in the early morning or late afternoon, when it emerges from the forest to feed. Sometimes, in the darker recesses of subtropical rainforests along the Queensland–New South Wales border, these shy marsupials can be spotted sunning themselves in morning sunlight that penetrates the forest canopy.

FEATURES: Greyish-brown fur takes on a red hue around the neck (leading to this species' moniker) and becomes a pale buff white on the underside. Small forepaws are usually held in close to the body. When hopping, the tail, which is held stiffly straight out behind, is a characteristic feature of this species.

DIET & HABITAT: Dense subtropical rainforests and eucalypt forests abutting well-grassed clearings form this pademelon's preferred habitat. Red-necked Pademelons are herbivorous, dining on grasses and shrubs. Populations can grow especially abundant in forests that adjoin agricultural or cleared land, where grass and fresh shoots are more plentiful.

BEHAVIOUR: Unlike many macropods, Red-necked Pademelons never use pentapedal locomotion. When moving slowly they are quadrupedal, simply dragging the tail behind rather than placing any weight on it. Tunnels are used as runways, but this species does not build nests, instead resting in shallow hollows in leaf litter on the forest floor. When feeding, pademelons often gather in small groups and use warning thumps and guttural growls to communicate approaching danger. Mothers use a high-pitched squeak and ongoing "click" noise to call to young.

BREEDING: At seventeen months of age, females can reproduce and breed year-round. Breeding peaks during spring and autumn.

THREATS: Dingoes, foxes, feral dogs, Wedge-tailed Eagles and habitat destruction are a concern.

Above: Small forepaws are held in close to the body. **Opposite, clockwise from top:** Red-necked Pademelons rarely move more than 100 m from the forest edge; Joey mortality increases once they leave the pouch; Keeping a lookout; Forepaws are used to manipulate food.

DIET: Grass and shrubs
LENGTH: HB 29–62 cm; T 27–51 cm
WEIGHT: 2.5–9.1 kg
STATUS: Secure

The Quokka

Unique among macropods, the stout, lovable Quokka is the only species in its genus and is regarded as a durable but superseded model of macropod. Although it resembles some of the other small wallaby species, its teeth, chromosomal make up, skull structure and blood proteins differ from those of all other macropods. The Quokka's dentition, especially, indicates that it is probably a survivor from the past — the result of evolution from earlier species of browsing macropods.

Quokkas browse on grass, leaves and succulent plants and may even climb low bushes to reach leaves when food is scarce. Rottnest Island has a seasonally harsh and arid environment; the Quokka's survival in this habitat depends on a small number of freshwater soaks, around which groups congregate and follow strict social rules.

Above: Rottnest Island. Before European settlement of Australia, the Quokka was relatively common in South-West Western Australia. By the 1960s, its range was confined to Rottnest Island and a few small areas around Perth. The Quokka is now carefully monitored in a small area of the South-West.

Quokka *Setonix brachyurus*

In 1696, Dutch navigator Willem de Vlamingh spied the Quokka on an island off the coast of present-day Western Australia. He mistakenly identified the marsupial as a large rat and named its home Rottenest in his native Dutch (or "Rat's Nest" in English). This species' island abode makes it easy to catch — hence the Quokka contributed much to the early research of macropods.

FEATURES: Superficially, the Quokka's major distinguishing features are its very short, scaly tail and compact body. Aside from the occasional dark stripe down its forehead, the Quokka has no definite body markings and is uniformly grey-brown on the back with a reddish tinge. The underbody is buff-coloured.

DIET & HABITAT: Although common on fox-free Rottnest Island, the rugged, often waterless terrain is not the Quokka's preferred environment and they seasonally suffer from nitrogen deficiency and lack of water. In the South-West, Quokkas seek out moist, heavily vegetated swamps, where they browse on foliage, grasses, sedges and succulent plants.

BEHAVIOUR: Quokkas are gregarious, diurnal animals, living in sociable groups that are dominated by the oldest males.

BREEDING: Births (a single joey) occur from January to March on Rottnest and all year on the mainland. Quokkas delay breeding if conditions are hot and dry.

THREATS: Foxes on the mainland; dehydration on Rottnest.

Left to right: The Quokka's sturdy forearms are used to "grip and strip" small branches; Quokkas are sociable macropods.

DIET: Grass, leaves, sedges and succulent plants, which may be low in nutrients

LENGTH: HB 40–54 cm; T 24.5–31 cm
WEIGHT: 2.7–4.2 kg
STATUS: Vulnerable

The Swamp Wallaby

The Swamp Wallaby is one of a kind — deserving of its own genus. In fact, as the sole member of the genus *Wallabia*, the Swamp Wallaby can be said to be the only true wallaby — if we take its scientific name literally.

Unlike other wallabies, it is more of a browser than a grazer and consequently has large premolar teeth with sharp cutting edges — features absent in the dentition of other wallabies. It is also more diurnal than other species, foraging or resting in thick undergrowth during the day and travelling to more open feeding grounds at night. The Swamp Wallaby's hopping gait is another distinction. When on the move, the head is held noticeably lower than that of other macropods and the tail extends rigidly out behind the body.

Obvious bodily differences are not the only reason the Swamp Wallaby lies in a league of its own — it is genetically distinct. Typical wallabies have sixteen chromosomes, but a male Swamp Wallaby has only eleven, and a female ten.

Above, top to bottom and right: The Swamp Wallaby's dark "mask" and red fur around the ears make it easy to tell apart from other species.

Swamp Wallaby *Wallabia bicolor*

The Swamp Wallaby's coarse, black-and-red-tinged fur probably contributed to its widespread distribution along Australia's east coast — saving it from the fur trade that threatened many other species.

FEATURES: The Swamp Wallaby's dark coat also makes this species easier to identify. Other prominent features are a black "mask" from each eye to the nostrils, black feet and forepaws and greyish-black inside the ears. A beige to white stripe runs along the jaw from the mouth to the ear. The thick, sometimes white-tipped, tail (which is held horizontal when hopping) is also a good indicator of this species.

DIET & HABITAT: The Swamp Wallaby occupies a wide range of habitats from tropical rainforest to woodlands, scrub, dry brigalow and heath. In heath, the Swamp Wallaby is easily the largest macropod species. It is a browser, able to digest plant varieties that are unpalatable or poisonous to other animals (including Bracken Fern and hemlock); however, the Swamp Wallaby seems to prefer the coarse, fibrous foliage of shrubs and grasses.

BEHAVIOUR: While usually solitary, small groups may be seen feeding together. The Swamp Wallaby is by nature a wary creature and is rarely observed in the wild.

BREEDING: When it comes to breeding, the Swamp Wallaby does things a little differently. Its gestation period is longer than its oestrous cycle, meaning that a second mating can take place (and result in a blastocyst in embryonic diapause) about eight days before the first embryo is born.

THREATS: Predation is by Dingoes, feral dogs and Wedge-tailed Eagles (mostly on joeys). Habitat destruction and land clearing are also a concern.

Above, top to bottom: Playfighting equips joeys with survival skills; A Dingo feasts on a Swamp Wallaby carcass.

DIET: Shrubs, ferns and grass across a range of habitats

LENGTH: HB 66.5–84.7 cm; T 64–86.2 cm
WEIGHT: 10.3–20.5 kg
STATUS: Vulnerable (SA); Secure elsewhere

Tree-kangaroos

Australia's two living tree-kangaroo species belong to an unusual group of macropods that have returned to their arboreal roots. All terrestrial macropods evolved from tree-climbing ancestors, but only tree-kangaroos reacquired features suited to life in the treetops. Tree-kangaroos have well-developed forelimbs roughly equal in length to the hindlimbs. The hindlimbs are shorter and broader than those of other macropods, and have textured soles for non-slip traction. The feet possess recurved claws to provide grip. A tree-kangaroo's hindlimbs can also move independently of each other — a trait common to tree-kangaroos and one that allows them better balance in the treetops.

Unlike the dentition of ground-dwelling macropods, a tree-kangaroo's teeth are designed for shearing off foliage rather than for grinding up high-fibre plant matter. Despite these developmental traits, their tails are not prehensile and their survival has probably more to do with a lack of predators or competitors in their habitat than with their climbing ability. When hopping along a branch or on the ground, the long tail is held behind the body as a counterbalance, rather than used for gripping objects. It is very rare to spot a glimpse of these nocturnal macropods during daylight hours. By night they forage in the canopy, seeking out tender leaves and fruit, but by day they sleep or rest high in the branches. Tree-kangaroos are sometimes seen feeding in small groups, but this is probably a product of food availability as they are mostly solitary animals.

Above: Bennett's Tree-kangaroo is the largest tree-living mammal in Australia.
Opposite: Lumholtz's Tree-kangaroo was first recorded by Norwegian scientist Carl Lumholtz in 1882.

Bennett's Tree-kangaroo *Dendrolagus bennettianus*

Unlike Lumholtz's Tree-kangaroo, Bennett's Tree-kangaroo seems to be equally at home in lowland rainforest and vine forest as it is in the high-altitude rainforests favoured by its arboreal relative. It is much larger than Lumholtz's Tree-kangaroo and is a speedy, powerful hopper when on the ground.

FEATURES: This species' dark reddish-brown body usually merges into fawn below. Although Bennett's Tree-kangaroo also has a grey-black mask from the nose to behind the eyes, it lacks the pale band running around the face and forehead in front of the ears. A major determining feature is the black patch at the base of the tail, which is not seen in Lumholtz's Tree-kangaroo.

DIET & HABITAT: Both highland and lowland rainforest, as well as vine forest, comprise suitable habitat for this species. In these environments it feeds largely on leaves, but has been known to eat fruit when available. Foliage from a number of plant species is favoured in lowland habitats, including *Pisonia* vine and *Platycerium* fern.

BEHAVIOUR: Bennett's Tree-kangaroo is a wary and secretive macropod, preferring to stay out of sight. Unusually, individuals occupy and defend a distinct home territory and adult males will fight viciously should an intruder enter their realm. Part of the territory of several females will usually fall within a male's territory, giving him a small harem of mates.

BREEDING: Little is known about the reproductive behaviour of this species, but it appears that breeding occurs year-round. Embryonic diapause may be used on occasion. Evidence suggests that young stay in the pouch for approximately nine months and leave to take up their own home range at about two years of age.

THREATS: Dingoes and large Scrub Pythons are this macropod's most formidable predators. Within its limited range, a persistent threat is degradation of habitat.

Above: Tree-kangaroos descend from trees tail-first. The front paws move alternately while the hindfeet slide against the trunk. About 2 m from the ground, the macropod pushes off, executes a mid-air twist and finishes with an upright landing.

DIET: Foliage, leaves and fruit of rainforest plant species

LENGTH: HB 69–75 cm; T 73–84 cm
WEIGHT: 8–13.7 kg
STATUS: Near Threatened (IUCN Red List); Rare (Qld)

Lumholtz's Tree-kangaroo *Dendrolagus lumholtzi*

Lumholtz's Tree-kangaroo probably evolved from ancestors that migrated to Australia from Melanesia. Before this species was brought to the attention of naturalist Carl Lumholtz in 1882, it was well known to Aborigines of the Atherton Tableland and Herbert River region, who knew it as "Boongarry", "Mabi" or "Muppie". Today, it remains the totem animal for an elder of the Ngadjon-Jii people, affording it special protection.

FEATURES: This species is smaller and darker in colour than Bennett's Tree-kangaroo, with a darker tail. Black shading on the face tends not to extend as far back towards the ears as for the other species; instead a paler grey-yellowish band runs around the face, behind the eyes and under the neck. There is usually more of a distinction between this species' darker, grey-black upper body and its buff, rufous-tinged underside than that seen in the Bennett's Tree-kangaroo. Its range, which does not overlap with Bennett's Tree-kangaroo, is another distinguishing factor.

DIET & HABITAT: Lumholtz's Tree-kangaroo is restricted to high-altitude rainforest between Cardwell Range and Mt Spurgeon in Queensland, where it feeds on leaves and fruit.

BEHAVIOUR: During daylight hours, this species sleeps in a crouched, slumped-over position in a tree fork or stretches out along a branch. They are able to move forwards and backwards, and even run, but will adopt a typical bipedal hop on broad branches. Despite being largely solitary, Lumholtz's Tree-kangaroos sometimes form small groups to feed, although males will fight in the presence of females. Males wooing a female in oestrus make gentle "clucking" noises and paw the female's head and neck.

BREEDING: Breeding is year-round and studies carried out in captivity indicate that a single joey is born and stays in the pouch for a little over 32 weeks. Each female will usually raise a single joey in a two year period.

THREATS: Clearing and logging of rainforest is a real threat to this species. Pythons, owls and feral cats will also prey on joeys and juvenile tree-kangaroos.

DIET: Fruit and foliage, including Ribbonwood and Wild Tobacco leaves

LENGTH: HB 52–65 cm; T 65.5–73.6 cm
WEIGHT: 5.1–8.6 kg
STATUS: Near Threatened (IUCN Red List); Rare (Qld)

Rock-wallabies

Rock-wallabies are the agile, able gymnasts of the macropod world, easily scaling the cliff faces and rocky escarpments of some of this country's steepest terrain. To navigate such rugged territory, these nimble rock-hoppers are equipped with special spring-loaded hindlimbs and thick, roughly textured soles (to enhance traction). Muscular tails help them balance and steer. When not deftly manoeuvring through and over rocks, rock-wallabies shelter in crevices, caves and overhangs; browse in clearings between boulders; keep watch; or descend from their rocky fortresses to graze.

Many rock-wallaby species are sociable, forming small colonies. In other, more solitary species, social interaction is limited to mating or shared feeding. The fossil record indicates that rock-wallaby species once migrated long distances from one area of suitable habitat to another. Over time, climate change, agriculture and development between habitats put a stop to these pilgrimages, isolating populations from each other. Gradually, members of the same species evolved different traits for their different habitats, and genetics put an end to interbreeding — creating distinct, highly specialised species.

Above: The Mareeba Rock-wallaby is one of seven rock-wallaby species that form a chain of similar species from Cape York to the New South Wales border.
Opposite: The Yellow-footed Rock-wallaby is one of Australia's most attractively patterned macropod species.

Short-eared Rock-wallaby *Petrogale brachyotis*

The Short-eared Rock-wallaby is one of Australia's least studied rock-wallaby species, despite its relative abundance in some areas of Arnhem Land in the Northern Territory. Although much larger than the Nabarlek and Monjon, with which it shares some of its range, scientific studies have shown these species to be its closest relatives.

FEATURES: This species varies widely in fur colour, markings and size, which sometimes makes it difficult to identify. Its most distinctive feature is its very short ears, which are mostly less than half the head's length. (Its larger size also helps differentiate it from the Monjon.) The fur is short and not particularly thick but is mostly grey-brown with silver speckles. The paws are black but the forearms vary from slightly rufous to whitish or beige. A dark dorsal stripe usually runs from the forehead to the middle of the spine. The somewhat short tail grows darker at the tip, where it is sparsely tufted.

DIET & HABITAT: Scattered populations occupy a wide range across northern Australia from the Kimberley to the Northern Territory–Queensland border. They prefer rocky territory in savanna woodland where rainfall is seasonal but usually around 600 mm per year. They also inhabit Groote Eylandt, Sir Edward Pellew Group, Wessel Islands and English Company Islands. They feed on grasses, sedges and seeds.

BEHAVIOUR: Short-eared Rock-wallabies are mostly nocturnal, but during overcast weather they may be seen basking on rocks. They are a gregarious species and are also very nimble, hopping away swiftly when disturbed.

BREEDING: Very little is known about the reproductive biology of this rock-wallaby species.

THREATS: Dingoes prey on this species, particularly on juveniles that have left the pouch, as do Olive Pythons. Land clearing and habitat destruction between patches of suitable habitat is also an ongoing threat.

Above: Forearms tucked and poised for escape, a Short-eared Rock-wallaby eyes an intruder.

DIET: Browses grasses, seeds and sedges

LENGTH: HB 40.5–55 cm; T 32–55 cm
WEIGHT: 2.2–5.6 kg
STATUS: Secure

Above, top to bottom: Male and female Short-eared Rock-wallaby; Short-eared Rock-wallaby with joey.

Monjon *Petrogale burbidgei*

The Monjon, Australia's smallest rock-wallaby species, is also sometimes commonly referred to as the Warabi. Its limited range in the rugged Kimberley region of Western Australia, and its superficial similarity to the Nabarlek, allowed it to avoid detection until 1978.

FEATURES: When seen up close, the Monjon has a distinct marbled appearance — its fawn or buff fur is speckled with darker olive and brown hues. The underside of the body is a pale fawn to ivory yellow and the shoulders, flanks and fur around the eye have a yellowish or reddish tinge. A pale, greyish, indistinct midline may be present on the forehead. The tail is fawn, becoming darker and quite thickly furred at the tip. Smaller ears and different habitat requirements distinguish it from the Nabarlek.

DIET & HABITAT: Inhabits low open woodland in high-rainfall coastal areas of the Kimberley where the sandstone outcrops of the King Leopold Range provide it with protection. It also makes its home on some islands in Western Australia's Bonaparte Archipelago. The Monjon is often associated with woodland that contains plenty of *Acacia, Terminalia, Eucalyptus* and *Ficus* species, among others.

BEHAVIOUR: This gregarious, mostly terrestrial species flees with remarkable speed if disturbed from its daytime hiding spaces under crevices or flattened between rocks.

BREEDING: Breeding probably occurs all year round and may peak during the wet season when food and water are more plentiful. Little else is known about its reproductive biology.

THREATS: Dingoes, feral cats and birds of prey (especially White-bellied Sea-Eagles) will make a meal of the small Monjon. Its isolated habitat largely protects it from habitat destruction by humans. Natural events such as drought and fire are threats.

Above: Figs form part of the Monjon's diet.
Opposite, top to bottom: The Monjon's tail is characteristically fluffy; Kimberley habitat.

DIET: Grass, leaves and some fruit, particularly figs

LENGTH: HB 30.6–35.3 cm; T 26.4–29 cm
WEIGHT: 960–1430 g
STATUS: Near Threatened (IUCN Red List)

Nabarlek *Petrogale concinna*

Sometimes also called the Little Rock-wallaby, the Nabarlek is the second-smallest of Australia's rock-wallaby species. It possesses a unique dental feature — it is the only marsupial equipped with an unlimited supply of grinding supernumerary molars. As the Nabarlek's teeth grind down, replacement molars erupt at the back of the jaw and move forwards to replace the worn-out teeth.

FEATURES: Fur is grey with a reddish tinge, particularly on the limbs, rump and tail. The last third of the tail is furred and darker brown in colour. Sometimes a dark stripe runs from the nose, over the forehead to the shoulder. On the cheeks, a white stripe runs below a darker patch of grey to blackish fur that extends from the nose to the eyes.

DIET & HABITAT: Nabarlek live among sandstone outcrops in the woodlands and coastal regions of the Northern Territory and northern Western Australia. Three likely subspecies include *Petrogale concinna canescens* (Arnhem Land and Groote Eylandt), *P. concinna monastria* (in coastal regions of the Kimberley) and *P. concinna concinna* (between Victoria and Mary Rivers). Its renewing supply of molars helps the Nabarlek grind down grasses and plants rich in abrasive silica, such as the fern *Marsilea crenata*.

BEHAVIOUR: Although timid around humans, these little rock-wallabies are sociable and gregarious with others of their kind. Nabarleks are generally nocturnal, but in the wet season may be seen basking on hot rocks in the early hours following dawn. They sometimes feed at dusk.

BREEDING: Nabarleks can most likely breed all year long, although most births occur during the wet season. The oestrous cycle is 32–35 days with a gestation period of 30–32 days. Joeys leave the pouch at around six months of age. Embryonic diapause has been observed in captive populations.

THREATS: White-bellied Sea-Eagles prey on this small species. Dry season fires are another natural hazard.

Above: The Nabarlek is a small, fine-featured rock-wallaby found in Australia's extreme north.

DIET: Plants and grasses, such as *Cyperus cuspidaus*, *Marsilea crenata* and *Eriachne* sp.

LENGTH: HB 29–35 cm; T 22–31 cm
WEIGHT: 1–1.7 kg
STATUS: Near Threatened (IUCN Red List & NT)

Above, top to bottom: The Nabarlek feature a small stature and fluffy, brush-like tail tip; The Nabarlek is an important native marsupial of the Arnhem Land escarpment.

Mareeba Rock-wallaby *Petrogale mareeba*

Although many of Queensland's rock-wallaby species are superficially very difficult to tell apart, DNA research has proved them to be distinct species. The Mareeba Rock-wallaby is one of these virtually indistinguishable types — almost impossible to discern from Sharman's and Allied Rock-wallabies. Although first recorded in 1974, it was only granted species status in 1992.

FEATURES: This small, mostly dark grey-brown rock-wallaby can vary considerably in colour depending on the rocky habitat in which it lives. Its underparts, forearms, hindlegs and tail base are often a pale buff colour with the extremities of the paws and feet taking on a darker, blackish hue. Usually there is a darker patch behind the forearms, although this is rarely as pronounced as in the very similar Godman's Rock-wallaby. A pale stripe is usually present on the cheek and sometimes a dark vertical stripe is seen between the ears on the forehead. The tail darkens towards the tip and often terminates in a white tip.

DIET & HABITAT: The Mareeba Rock-wallaby dwells in open forest surrounded by granite and basalt outcrops in a parcel of land in north-east Queensland, from Mt Carbine to Mt Garnet. It survives on grass, leaves, forbs and low shrubs.

BEHAVIOUR: Mareeba Rock-wallabies are sociable macropods and form colonies of up to 100 individuals. Mutual grooming is also a part of their social structure. They are mostly nocturnal, sheltering under rocky overhangs or in caves by day.

BREEDING: Mareeba Rock-wallabies form lasting pair bonds. Usually, only one joey is produced at a time and stays in the pouch for around 26 weeks. Once it vacates the pouch, but before it is weaned, it is often left in rocky caves while the mother goes out to feed.

THREATS: Dingoes, foxes and clearing of land between patches of habitat are the Mareeba Rock-wallaby's greatest threats.

Above: Mareeba Rock-wallaby habitat.
Opposite, top to bottom: A joey on the verge of being fully weaned enjoys a suckle; A Mareeba Rock-wallaby effortlessly negotiates its stone habitat.

DIET: Grass, shrubs, forbs and leaves

LENGTH: HB 42.5–54.8 cm; T 41.5–53 cm
WEIGHT: 3.8–4.5 kg
STATUS: Secure

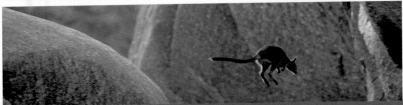

Godman's Rock-wallaby *Petrogale godmani*

This species is endemic to Queensland and was first described by Oldfield Thomas in 1923 working from a specimen collected at Black Mountain in 1922. Its present range extends from near Mt Carbine to Bathurst Head. It was once quite common in the Black Mountain region south-west of Cooktown, but is now rarely spotted there.

FEATURES: Godman's Rock-wallaby varies in colour depending on the rock colour of its environment. However, for the most part it remains similar to the Allied, Unadorned, Sharman's and Mareeba Rock-wallaby species in appearance. A slight distinction is that its forepaws are often a little more buff-coloured, as is the base and first third of the tail. Studies of its chromosomes link it closely to the Allied Rock-wallaby, but analysis of its blood and tissue proteins indicate that the Cape York Rock-wallaby is also a close relative.

DIET & HABITAT: Godman's Rock-wallaby finds shelter in steep cliffs, outcrops and boulders set among areas of open woodland on Queensland's north-east coast around the towns of Cooktown and Laura.

BEHAVIOUR: Little information is available on behavioural aspects of this species. Like most rock-wallaby species it probably forms colonies and is nocturnal — seeking shelter from the heat in caves, crevices or overhangs of its rocky territory.

BREEDING: Limited research is available on the breeding and reproductive habits of this species. What is known is that fertile hybrids have been born in southern areas of its range near Mitchell River, where this species' distribution overlaps with that of the Mareeba Rock-wallaby. There is also evidence of Godman's Rock-wallaby genes in some Mareeba Rock-wallabies south of the known hybrid range.

THREATS: Like most rock-wallaby species, it is threatened by climate change and habitat destruction and preyed upon by Dingoes, feral cats and Wedge-tailed Eagles.

Above: Black Mountain, near Cooktown, where the type specimen of Godman's Rock-wallaby was found. This species was once common and abundant around Black Mountain but is now rarely seen in the area.

DIET: Grass, leaves and shrubs

LENGTH: HB 49.5–57 cm; T 48–64 cm
WEIGHT: 4.3–5.2 kg
STATUS: Secure

Herbert's Rock-wallaby *Petrogale herberti*

This widespread, common macropod of rocky climes along the Great Dividing Range in Queensland's south-east was long thought to be a subspecies of the Unadorned Rock-wallaby or Brush-tailed Rock-wallaby, despite being described as a species in 1926. DNA research, however, has seen Herbert's Rock-wallaby reinstated as a distinct species.

FEATURES: This species is best identified by a study of its chromosomes. In the field, only subtle differences exist between it and similar rock-wallaby species. One feature is that the white stripe running from front shoulder to flank appears stronger on this species, and a marked black dorsal stripe extends over the forehead to the shoulder. The brush on the tail tip is also less bushy than that of the Brush-tailed Rock-wallaby.

DIET & HABITAT: Occupies quite a large area stretching from Nanango in Queensland northwards to Rockhampton and inland as far as Clermont and Rubyvale. Like all rock-wallabies, its diet consists mainly of grasses, leaves and shrubs.

BEHAVIOUR: The sociable and gregarious Herbert's Rock-wallaby forms localised colonies. Parts of its range overlap with the distribution of the Brush-tailed Rock-wallaby and the two species coexist harmoniously, making accurate identification more difficult.

BREEDING: This species is so similar to the Brush-tailed Rock-wallaby that fertile hybrids can result from cross-breeding, yet chromosomal and blood and tissue protein differences exist. Females are able to breed throughout the year and practice embryonic diapause when necessary, keeping an embryo-in-waiting in the pouch until more food and water are available.

THREATS: Dingoes, foxes, Wedge-tailed Eagles and clearing of land between suitable "safe havens" of habitat threaten most of Queensland's rock-wallaby species.

Above: Herbert's Rock-wallaby is slightly larger than the similar Unadorned Rock-wallaby and smaller than the Brush-tailed Rock-wallaby.

DIET: Grasses, leaves and shrubs

LENGTH: HB 47–61.5 cm; T 51–66 cm
WEIGHT: 3.7–6.7 kg
STATUS: Secure

Allied Rock-wallaby *Petrogale assimilis*

The Allied Rock-wallaby was first discovered on Palm Island and recorded by Edward Ramsay (then curator of the Australian Museum) in 1877. Palm Island and Magnetic Island, along with a swathe of country inland from Townsville to Hughenden, still form this species' distribution range today.

FEATURES: This species is remarkably indistinct in feature (similar to the related Unadorned Rock-wallaby) and lacks prominent markings. Its fur colour varies to suit different rock types, although it is mostly grey-brown on the body with yellowish-brown underparts, forearms and hindlimbs. Some sport a pale cheek stripe and a reddish patch at the base of the tail. The tail grows darker towards the tip and displays a slight brush.

DIET & HABITAT: The areas this species inhabits are often seasonally harsh and unpredictable, alternating between drought and deluge. To counter this, the Allied Rock-wallaby is a more opportunistic feeder than some other species and will browse on a range of forbs (herbaceous plants), shrubs, fruits, leaves, seeds and flowers. After rain, fresh green pick is a favoured food.

BEHAVIOUR: Allied Rock-wallabies form colonies in which a hierarchy is established based on age. Mature males and females pair bond, groom each other and forage together. These rock-wallabies are mostly nocturnal.

BREEDING: Research conducted on captive specimens has shown that the oestrous cycle and gestation period for this species are about the same length — 30–32 days. Births occur year-round and a single joey is produced each time, remaining in the pouch for 6–7 months. Both males and females reach sexual maturity and start breeding at approximately eighteen months.

THREATS: Dingoes, foxes and Wedge-tailed Eagles prey on this species, particularly juveniles that have left the pouch. Land clearing and habitat destruction between patches of suitable habitat are also an ongoing threat to its secure status.

Above: This Allied Rock-wallaby appears to be contemplating a leap. Note how the tail counterbalances the body weight.

DIET: Forbs, shrubs, leaves, seeds, shoots and flowers

LENGTH: HB 44.5–59 cm; T 40.9–55 cm
WEIGHT: 4.3–4.7 kg
STATUS: Secure

Unadorned Rock-wallaby *Petrogale inornata*

In 1842, John Gould classified this wallaby as the species Petrogale inornata *due to the absence of obvious body markings. Its lack of adornment makes it somewhat difficult to determine from the Sharman's Rock-wallaby, Mareeba Rock-wallaby or Allied Rock-wallaby.*

FEATURES: Because it lacks prominent markings, this species can easily be confused with others. Fur colour varies, allowing this species to blend in with various types of rock. Usually it is grey-brown on the upper body and a paler, sandy colour on the forearms, hindlimbs and underside. Sometimes a pale cheek stripe and a darker stripe down the centre of the forehead are observed — usually in the south of this species' range where individuals are slightly more marked. The tail grows darker towards the end and has a sparse brush at the tip; some have a white tip.

DIET & HABITAT: Unadorned Rock-wallabies inhabit rocky habitats in open woodland or vine thickets from Rockhampton in Queensland's north and along the central coast to Home Hill. They are also present on some of the Whitsunday Islands. Like all rock-wallabies, Unadorned Rock-wallabies are herbivores, feeding primarily on grass, and the foliage of trees and various shrubs.

BEHAVIOUR: Individuals form colonies and adult males and females pair bond. Shared grooming and shared foraging are good indicators of a pair.

BREEDING: This species appears capable of reproducing all year with no discernible breeding season. Both males and females become sexually mature at approximately eighteen months of age. A single joey is born after a gestation period of 30–32 days and will remain in the pouch for 6–7 months.

THREATS: Dingoes, foxes and Wedge-tailed Eagles prey on this species, particularly on juveniles that have left the pouch. Land clearing of suitable protective vegetation between patches of preferred habitat is an ongoing threat to most rock-wallaby species.

Above: Unadorned Rock-wallabies resting. Many rock-wallabies sit on their tails while grooming.

DIET: Grass, leaves and shrubs

LENGTH: HB 45.4–57 cm; T 43–64 cm
WEIGHT: 3.1–5.6 kg
STATUS: Secure

Sharman's Rock-wallaby *Petrogale sharmani*

Sharman's Rock-wallaby (also known as Mt Claro Rock-wallaby), was only granted species status in 1992. This species' name pays tribute to Professor G.B. Sharman, who conducted extensive scientific research into the genus Petrogale *and contributed much to the study of marsupial biology in general.*

FEATURES: Although it has a different number and shape of chromosomes, Sharman's Rock-wallaby is, to the naked eye, almost indistinguishable from the Mareeba Rock-wallaby — a species that competes with the Sharman's Rock-wallaby in the north of its range. Mostly grey-brown on the upper body, fur becomes a pale yellowish-brown on the forearms, hindlimbs and underparts. Sometimes there is also a buff-coloured patch at the tail's base. The tail becomes darker towards the end and the tip terminates in a slight brush. In some, a stripe down the centre of the forehead may be seen.

DIET & HABITAT: Hemmed in by the Allied Rock-wallaby to the south and west and the Mareeba Rock-wallaby to the north, Sharman's Rock-wallaby occupies only a limited territory of around 200,000 ha in the Seaview and Coane Ranges near Ingham, Queensland. Their diet remains consistent with that of all other rock-wallabies and they feed largely on grass, leaves and shrubs.

BEHAVIOUR: Although mostly nocturnal in the summer months, Sharman's Rock-wallaby is sometimes active at dusk in the winter months and may also be seen basking on rocks in the early morning and late afternoon.

BREEDING: Offspring are produced throughout the year after a gestation period of 30–32 days. The oestrous cycle is the same length as the gestation period. Joeys stay in the pouch for around 6–7 months.

THREATS: Due to its limited distribution, this species is particularly vulnerable to habitat destruction and environmental changes. Introduced predators, such as feral cats and dogs, and competitors such as cattle and goats also impact its population.

Above: Although primarily a nocturnal species, Sharman's Rock-wallabies may be spied in the early morning and late afternoon during winter.

DIET: Grass and foliage of shrubs and trees

LENGTH: HB 45.5–53 cm; T 43.5–53.2 cm
WEIGHT: 4.1–4.4 kg
STATUS: Near Threatened (IUCN Red List); Rare (Qld)

Cape York Rock-wallaby *Petrogale coenensis*

Inhabiting only a small area near the town of Coen on Cape York Peninsula, this is one of Queensland's rarest rock-wallaby species. It was initially thought to be a northern race of Godman's Rock-wallaby, but was reclassified as a chromosomally distinct species in 1992.

FEATURES: Visually, the Cape York Rock-wallaby appears very similar to Godman's Rock-wallaby, with the same grey-brown colouring on the body and sandy, buff-coloured underparts, hindlimbs, forearms and tail base. The last third of the tail is a greyish-silver in most specimens and terminates in a characteristic silver-coloured tail tip. Some recorded specimens appear to differ considerably. One specimen had short, almost white fur on its underside; another appeared to have mottled, purplish spots on the belly and towards the flanks.

DIET & HABITAT: Prefers rocky slopes, screes, boulders and outcrops among open woodland. Some areas of its range also contain vine thicket. Like other rock-wallabies, it survives on a diet of grass, leaves and shrubs.

BEHAVIOUR: Very little is known about the behaviour of this species. The Cape York Rock-wallaby appears to be quite rarely seen, even within its range.

BREEDING: Further research is required into the reproductive behaviour of this species, but it is likely that reproduction is relatively similar to that of other north Queensland rock-wallaby species.

THREATS: Due to its limited distribution, and apparently scattered populations within its range, the Cape York Rock-wallaby is particularly vulnerable to habitat destruction and environmental changes. Its current status is described as Low Risk (Near Threatened) on the World Conservation Union's IUCN Red List and more information is needed to ascertain numbers. Introduced predators, such as feral cats and wild dogs, and competitors such as cattle and goats could also impact on this species' populations.

Above: Typical Cape York Rock-wallaby habitat.

DIET: Grass, leaves and shrubs

LENGTH: HB 44–56.4 cm; T 47–54 cm
WEIGHT: 4–5 kg
STATUS: Near Threatened (IUCN Red List); Rare (Qld)

Black-footed Rock-wallaby *Petrogale lateralis*

Perhaps confusingly, it is the sole of this wallaby's foot that is black, rarely the entire back foot, which is often a rich dark-brown with a slight reddish tinge. This rock-wallaby species includes three recognised subspecies — Petrogale lateralis lateralis, P. lateralis hacketti and P. lateralis pearsoni — and two other chromosomally distinct races (the West Kimberley Race and the MacDonnell Ranges Race).

FEATURES: The Black-footed Rock-wallaby varies considerably depending on the subspecies or race. Often, this species has fluffy, dark grey-brown fur above with paler brown forearms, hindlegs and tail base. The toes and forepaws are brownish to black and the soles of its feet are black. A distinct white cheek stripe, and a brown stripe down the centre of the forehead to beyond the shoulders is usually present. A white side stripe extends from the shoulder to the flank above a dark-brown/blackish underarm patch. The tail is dark brown to black with a slight brush at the tip.

DIET & HABITAT: Populations and subspecies are recorded in a number of rocky habitats in the Northern Territory, South Australia and Western Australia. They also inhabit some Western Australian islands and the Investigator Group in South Australia. Its diet is mostly grass, leaves and fruit.

BEHAVIOUR: John Gilbert first recorded the behaviour of this species in the 1840s, writing that it was a "remarkably shy and wary animal". He also noted that it rarely ventured more than a few hundred yards from rocky shelter. It may be seen basking during the day but is mostly nocturnal. Males and females in a colony have separate hierarchies.

BREEDING: Little research is available, but current studies indicate that they probably breed all year round and exhibit embryonic diapause.

THREATS: Subspecies have declined across much of their range, probably from predation by the Red Fox.

Above: Pearson Island Rock-wallaby — a subspecies of the Black-footed Rock-wallaby.

DIET: Grasses, leaves and some fruits

LENGTH: HB 45–52.1 cm; T 50.7–59.7 cm
WEIGHT: 2.8–4.5 kg
STATUS: Vulnerable in parts of its range

Above: The Black-footed Rock-wallaby includes three subspecies and two distinct races.

Brush-tailed Rock-wallaby *Petrogale penicillata*

This once widespread species was the first rock-wallaby to be classified by European scientists, being given the name Kangurus penicillatus *in 1825. In 1837, its scientific name was changed to* Petrogale pencillata. *Its abundance made it an easy target for the fur trade — in just a year, 92,590 skins were sold by a single Sydney company.*

FEATURES: This rock-wallaby is similar in appearance to many other species (especially Herbert's Rock-wallaby). A distinguishing feature is the prominent brush on the tail. Its brownish, fluffy fur on the upper body is paler beneath, usually with a dark patch in the underarm. The face has a white to beige cheek stripe and a black stripe from the back of the head extending vertically down the forehead to the nose bridge. The paws are dark brown and usually darken towards the tip.

DIET & HABITAT: Grass and forbs are a major dietary component, although Brush-tailed Rock-wallabies will also feed on flowers, fruit and seeds. It occupies numerous habitats where protective rocky outcrops are found, including wet and dry sclerophyll forests, rainforest edges, open woodland and semi-arid regions. Usually, they make their homes on rocks that face north, allowing them to bask during the early morning and afternoon. This species has been introduced to New Zealand and Hawaii.

BEHAVIOUR: This extremely nimble species is mostly nocturnal, but is often seen basking in the sun by day. They are sociable, forming small colonies.

BREEDING: Female Brush-tailed Rock-wallabies are mature at 1–2 years and are then fertile all year. Embryonic diapause is used in times of drought. Unusually, the young of this species are weaned as soon as they exit the pouch but continue to stay with the mother for another few months.

THREATS: Competing with introduced species such as rabbits and goats has been detrimental. Dingoes and foxes are major predators.

Above: In some regions of its range near Nanango, Queensland, the Brush-tailed Rock-wallaby is able to interbreed with Herbert's Rock-wallaby to produce fertile hybrid offspring. However, this occurrence is not common.

DIET: Grass, forbs, seeds, fruit and flowers

LENGTH: HB 51–58.6 cm; T 50–70 cm
WEIGHT: 4.9–10.9 kg
STATUS: Vulnerable (IUCN Red List & Cmwth)

Above: To catch the morning and afternoon sun, Brush-tailed Rock-wallabies regularly make their homes with a north-facing aspect.

Yellow-footed Rock-wallaby *Petrogale xanthopus*

The Yellow-footed Rock-wallaby is Australia's largest rock-wallaby species. It is also one of the most attractively patterned rock-wallabies. Because of this, it was exploited for fur and skins in the late 19th century.

FEATURES: As its name implies, the Yellow-footed Rock-wallaby does indeed have yellow feet, but its tail is probably its most defining feature. Irregular dark bands run around the second half of the orange-brown tail — a feature not found in other rock-wallaby species. On the body, fur is light brown or grey with a yellowish tinge. The hindlimbs, forelimbs and part of the shoulder are yellow. The ears, too, are fringed with yellow fur. A buff to white side stripe and similarly coloured cheek stripe are also present.

DIET & HABITAT: Open woodland in semi-arid areas, close to rocky structures and permanent water, comprises this species' preferred habitat. Grass, forbs and foliage from trees and shrubs dominate its diet. These rock-wallabies are particularly common in the Flinders Ranges, South Australia, but also have a substantial population in rocky country south-west of Blackall in Queensland. A smaller population exists in north-western New South Wales.

BEHAVIOUR: The Yellow-footed Rock-wallaby is gregarious, forming colonies of up to 100 individuals. Each rock-wallaby has a home range of up to 200 ha, which overlaps that of other individuals in the colony.

BREEDING: Both males and females reach sexual maturity at around eighteen months of age. Females come into oestrus every 32–37 days and give birth to a single joey after 31–33 days of gestation. The young joey stays in the pouch for around 26 weeks.

THREATS: Wedge-tailed Eagles and Dingoes prey on Yellow-footed Rock-wallabies. Competition with introduced herbivores is also a real threat. Sheep and goats degrade the habitat favoured by Yellow-footed Rock-wallabies and compete with this species for food.

Above: The highly distinctive tail and yellow colouring on the feet and forepaws make this species one of the most vividly coloured and attractive of the rock-wallabies.

DIET: Grass, leaves and forbs

LENGTH: HB 48–65 cm; T 56.5–70 cm
WEIGHT: 6–11 kg
STATUS: Near Threatened (IUCN); Vulnerable (SA); Endangered (NSW)

Above: Australia's largest rock-wallaby species, the Yellow-footed Rock-wallaby creates a striking impression.

Proserpine Rock-wallaby *Petrogale persephone*

The endangered Proserpine Rock-wallaby is the third-largest species in the genus Petrogale. *Males are especially large, growing up to 60% bigger than females. Strangely, for a species that is distributed relatively close to a well-settled area, this species was overlooked until 1976, when it was finally brought to the attention of the scientific community.*

FEATURES: The Proserpine Rock-wallaby is a highly distinctive species, easily distinguished from other rock-wallabies. Its dark grey fur often has a slightly purple or mauve tinge on the rump and neck. The forearms, back of the hindlimbs, ears and much of the tail are a noticeable rufous-orange. A white to beige strip extends from the upper lip to below the ear. The chin is white and the undersurface a pale yellow, which may take a darker shade in some individuals. Digits are black, as is the latter part of the tail, although the brush-like tip is usually yellowish to white.

DIET & HABITAT: Of all rock-wallaby species, this species has the smallest distribution. Roughly 26 populations are spread through rocky habitat in open woodland and dry vine forest around Proserpine and on some of the Whitsunday Islands. These macropods prefer areas with a dense understorey of grass — which forms part of their diet, together with broad-leaved plants, coastal shrubs and pandanus.

BEHAVIOUR: These rock-wallabies are sometimes observed leaping up the branches of sloping trees, which has led to some people mistakenly calling them tree-kangaroos. They are partly diurnal and can be seen sunning during the day when the weather is cool.

BREEDING: The female Proserpine Rock-wallaby has an oestrous cycle of 33–35 days. Gestation takes almost the same amount of time (33–34 days), and the doe usually mates again hours after giving birth. Young remain in the pouch for about 30 weeks and are weaned 17–18 weeks after leaving the pouch.

THREATS: The Proserpine Rock-wallaby's extremely limited range in an area of heavy human settlement makes it extremely susceptible to habitat damage. Motor vehicles account for some deaths in the Airlie Beach area.

Left and opposite: The distinctive colours and markings of this species, and its finite range, make it easily identifiable.

DIET: Grasses, broad-leaved plants, pandanus, coastal shrubs

LENGTH: HB 50.1–64 cm; T 51.5–67.6 cm
WEIGHT: 4.1–8.8 kg
STATUS: Endangered (IUCN Red List & Cmwth)

Purple-necked Rock-wallaby   *Petrogale purpureicollis*

Although first classified as a species in 1924, the Purple-necked Rock-wallaby was for many years believed to be a subspecies of the Black-footed Rock-wallaby and was referred to as Petrogale lateralis purpureicollis. It has since been reclassified as a separate species based on genetic differences. Two of its eleven chromosomes are shaped differently to those of the Black-footed Rock-wallaby.

FEATURES: A highly distinctive species due to the purple tint of hair on this rock-wallaby's head, neck and shoulders. The intensity of the purple pigmentation changes depending on the time of the year and the specimen. The colour is exuded by this species' skin and washes out in water. In some seasons, fur on the face, neck and shoulders takes on a bright red to purple hue. The rest of the body is a light grey to brown with a black muzzle and a dark brown dorsal stripe down the centre of the forehead. A pale cheek stripe is often present below the eyes. The tail is silver to sandy brown, darker at the end and less brushy than that of the Black-footed Rock-wallaby.

DIET & HABITAT: Cliffs, gorges, outcrops and rocky country set in dry acacia woodlands, usually with a spinifex grass understorey. This species exists in disparate Queensland populations from Winton in the east to Lawn Hill Gorge in the north. Water availability is crucial, so most colonies form in areas where there is near-permanent water in deep holes throughout the year.

BEHAVIOUR: This rock-wallaby is a sociable species, forming colonies close to permanent waterholes.

BREEDING: The oestrous cycle (36–38 days) is slightly longer than the gestation period (33–35 days) and a female comes into oestrus and mates immediately after giving birth. Pouch life is around six months.

THREATS: Birds of prey and habitat destruction are threats.

Above: Because the purple pigmentation seeps from the skin and washes off (often disappearing within 25 hours), scientists initially thought that it must have come from the animals rubbing against rocks or tree bark. Males usually have brighter purple colouring than females.

DIET: Grass

LENGTH: HB 49–61 cm; T 45–61 cm
WEIGHT: 4.7–7.1 kg
STATUS: Secure

Rothschild's Rock-wallaby *Petrogale rothschildi*

In 1901, the first specimen of this large Western Australian rock-wallaby was sent to the Curator of Mammals at the British Museum. He classified it and named it Rothschild's Rock-wallaby after the Honourable Walter Rothschild, the patron of the expedition during which the specimen was found.

FEATURES: Pale greyish to golden fur on the upper body becomes a dull brown on the underside. The top of the head and ears are a dark brown in contrast with lighter coloured fur on the throat and neck. A pale cheek stripe is present but there is no dorsal stripe. The long tail becomes darker at the tip. In some individuals, the fur around the neck, shoulders and flanks may have a mauve tinge.

DIET & HABITAT: This species is associated with shrub and grass-steppe habitat around the Hamersley Range. It also inhabits some islands in the Dampier Archipelago, particularly those from which the introduced Red Fox is absent. Grasses, especially those growing in sandy soils between rocky outcrops, comprise most of its diet.

BEHAVIOUR: By day, Rothschild's Rock-wallaby rests in caves and crevices. It comes out to forage at night.

BREEDING: Little information is available on the reproductive biology of this species.

THREATS: Foxes have proven to be a significant threat to Rothschild's Rock-wallaby. The species was formerly recorded on Lewis Island in the Dampier Archipelago, but has now vanished from this range. It exists on the outer islands of Dampier Archipelago where the fox is not present. Rothschild's Rock-wallaby has experienced a general decline on the Western Australian mainland, probably as a result of predation by foxes.

Above: Rothschild's Rock-wallaby is common on four islands in the Dampier Archipelago off the northern Western Australian coast, including Rosemary, Dolphin, Enderby and West Lewis Islands. It was reintroduced to West Lewis Island from Enderby Island stock.

DIET: Grasses and vegetation

LENGTH: HB 46.3–59.2 cm; T 53.9–70.4 cm
WEIGHT: 3.7–6.6 kg
STATUS: Secure

Nailtail Wallabies

Shy, dainty nailtail wallabies take their name from a hard, nail-like spur on the end of the tail. Although highly distinctive, the spur appears to have no particular function. Three nailtail species once inhabited Australia's semi-arid shrubland and grassy woodlands. Today, the Crescent Nailtail Wallaby is extinct. The Bridled Nailtail Wallaby was presumed extinct for some time until, in 1973, a cattle farmer in Central Queensland recognised an illustration of the species as being the same as those that visited his property. The property, near Dingo in Queensland, was later acquired and became Taunton Scientific Reserve, where this species is now protected. Only the Northern Nailtail Wallaby remains common and is widely distributed across the Top End.

Nailtail wallabies have a curious style of locomotion. When hopping bipedally, they hold their forearms out in front of the body, rotating them in small circles. This has given this group of wallabies the colloquial name "organ grinders".

Above: Before European settlement, the now-extinct Crescent Nailtail Wallaby occupied semi-arid and arid habitats in Central and Western Australia. **Right:** The purpose of the "nail" has mystified scientists, but these photographs taken in 2007 appear to show it being used as a weapon. **Opposite:** The endangered Bridled Nailtail Wallaby has been bred in captivity at Scotia Wildlife Sanctuary.

Bridled Nailtail Wallaby *Onychogalea fraenata*

That this endangered species exists today is little more than happy chance. For years it fell victim to the fur trade and was considered extinct until it was "rediscovered" on an 11,400 ha property near Dingo, Queensland, now protected as Taunton Scientific Reserve. Its numbers have increased tenfold since cattle were excluded from its habitat.

FEATURES: The Bridled Nailtail is one of Australia's prettiest wallaby species. The upper body is a flecked grey, over which a prominent "bridle" of white runs from the shoulders around the armpits to meet up with a white-yellow underbody. White also runs under the chin and down the neck. A black strip runs from the nostrils to the eyes and is further delineated by a white strip underneath, which extends to the white-fringed ears. A black patch in the armpits and black dorsal stripe down the forehead complete this wallaby's adornment. The squat, yellowish-grey tail terminates in the characteristic nail.

DIET & HABITAT: Once widespread from Lake Hindmarsh, Victoria, to Charters Towers, Queensland, the Bridled Nailtail is now confined to acacia woodland with a grassy understorey in Taunton Scientific Reserve. It has also been reintroduced to Idalia National Park in Queensland. Bridled Nailtail Wallabies browse on some shrubs, and graze on grasses, drought-resistant shrubs and forbs.

BEHAVIOUR: Nailtail wallabies are more diurnal than most macropods and do not form colonies like rock-wallabies. When chased, the Bridled Nailtail typically attempts to hide, rather than outrun opponents, which makes it easy prey for Dingoes and feral predators.

BREEDING: Females breed all year, giving birth to a single joey that stays in the pouch for around eighteen weeks.

THREATS: Dingoes and feral cats prey on Bridled Nailtail Wallabies. Competition with introduced herbivores probably greatly contributed to this species' decline.

Top: Bridled Nailtail Wallaby. **Opposite, clockwise from top left:** After a cold night, individuals warm up directly in the sun; On the hop; They retreat to the shade in the midday heat, feeding later on when it is cooler.

DIET: Forbs, chenopods and soft-leaved grass

LENGTH: HB 43–70 cm; T 36–54 cm
WEIGHT: 4–8 kg
STATUS: Endangered (IUCN Red List & Cmwth)

Northern Nailtail Wallaby *Onychogalea unguifera*

The Northern Nailtail Wallaby was first recorded in 1838, on one of the HMS Beagle's *expeditions. This sandy yellow, sizeable wallaby is the largest of Australia's nailtail species. It is also the most widely distributed, extending from inland grassy plains to coastal floodplains across Australia's far north.*

FEATURES: Ginger fur on the upper body becomes paler on the head and neck and fades to a creamy white on the underbody. Circles of whiter fur can also be seen around the eyes. The long, pale grey ears are also fringed with white on the inside. A dark stripe runs from the shoulders, along the spine to the base of the tail, which is a pale beige to cream. The tail darkens towards the tip and terminates in the distinctive "nail".

DIET & HABITAT: Open woodlands with an understorey of tussock grass, as well as tall shrublands and coastal plains of tea-tree are preferred habitat. They seem to especially favour grasslands growing on blacksoil plains, like the Northern Territory's Barkly Tableland. They are fussy eaters, rarely eating shrub foliage but seeking out herbs, succulents, fresh green pick and fruit.

BEHAVIOUR: While they are mostly solitary, up to four individuals sometimes feed together, mostly between dawn and dusk. By day they rest in shallow hollows dug under low bushes — Conkerberry bushes, particularly, appear to be a favourite. When alarmed, this species emits a "wut, wut" noise.

BREEDING: Females are probably fertile all year and give birth to one offspring. Little else is known about the Northern Nailtail Wallaby's reproductive biology.

THREATS: This species has endured European settlement much longer than others that share its genus. Unlike the Bridled Nailtail Wallaby, there is little to suggest that sharing its environment with cattle has been detrimental, but very little of its range is protected as national park or reserves.

Above: A female Northern Nailtail Wallaby scratches an itch on her nail-tipped tail.

DIET: Succulents, herbs, forbs and some grasses

LENGTH: HB 49–69 cm; T 60–73 cm
WEIGHT: 4.5–9 kg
STATUS: Near Threatened (NT); Secure elsewhere

Clockwise from top: Males can grow up to 9 kg; Some individuals have speckles on the tail and face, although this is uncommon; A doe cleans her pouch before each new arrival and keeps it clean for the duration of the joey's pouch life.

Wallabies, Wallaroos & Kangaroos

Kangaroos, wallaroos and wallabies in the genus *Macropus* are Australia's most widespread, well-known and celebrated marsupials. Some, particularly the Red Kangaroo and Eastern Grey Kangaroo, have become synonymous with Australia and are well represented in wildlife parks and reserves around the continent.

Of all the macropod genera, *Macropus* is probably the most diverse and successful. Species belonging to this genus range in size from the small Parma Wallaby to the imposing male Red Kangaroo. Since European settlement, several of the larger, open-range species have increased in both number and distribution thanks to the additional food and groundwater provided by agriculture. Only one species of this genus has suffered extinction — the attractive, dainty Toolache Wallaby, which was hunted for its exceptionally beautiful pelt.

Most of these macropods are sociable and live in large groups known as mobs. Each mob usually consists of a dominant male (or buck) and several females (does) with their young. Even solitary species often feed in pairs or gather in small groups at prime, communal feeding sites.

Above: Eastern Grey Kangaroos gather in mobs with a strict social hierarchy.
Opposite, top to bottom: A Red-necked Wallaby mother and juvenile; Common Wallaroo; Eastern Grey Kangaroo and joey.

DISTINCTIONS BETWEEN THE THREE GROUPS ARE MADE ON THE BASIS OF SIZE AND HABITAT.

WALLABIES are smaller, generally weighing less than 25 kg (although males of some species may grow slightly larger). Due to their smaller stature, wallabies tend to rely on the protective shelter of habitats with dense undergrowth.

WALLAROOS are macropods that prefer mountainous, hilly country to the expansive undulating plains favoured by larger kangaroo species. They have large, bare, black noses and a distinctive upright stance with shoulders thrown back, elbows tucked into the sides and wrists raised.

KANGAROOS are the largest grazing macropods, with some individuals weighing upwards of 80 kg. Kangaroos are probably the most commonly seen macropods and include the well-known Eastern Grey, Western Grey and Red Kangaroo.

Agile Wallaby *Macropus agilis*

The Agile Wallaby is found in New Guinea as well as across Australia's tropical north, where it is widespread and common. Small populations also exist on Stradbroke and Russell Islands in South-East Queensland. Such is its abundance in the north that it has in the past been declared a pest in some agricultural areas, particularly the canefields of northern Queensland.

FEATURES: The Agile Wallaby has a pale, sandy colouring that grows even paler on the underparts and the forelimbs and hindlimbs. The ears are long, pointed and have a black fringe. A black dorsal stripe is usually seen on the forehead. The tip of the tail is also black. A distinct white stripe marks the thigh.

DIET & HABITAT: The Agile Wallaby prefers open riparian habitat and seasonally flooded grasslands. In the Northern Territory it is also common in sand dune country close to the coast. It feeds mostly on grasses, but may also excavate for the roots of some plant species, especially Ribbon Grass, and has been known to eat the leaves of Coolibah trees as well as fruit from figs and the Leichhardt Tree. Following fire and flood, it feasts on the fresh shoots of replenishing grass species.

BEHAVIOUR: Mobs of up to ten individuals live together and sometimes join with other groups to feed on good pastures. Males grow larger than females and each group is dominated by an "alpha" buck. When danger threatens, members of the group will thump their hindfeet to warn others.

BREEDING: Females reproduce year-round and a single joey is born after a 30-day gestation. Females mate again shortly after giving birth, which results in a quiescent blastocyst that begins to develop shortly before the pouch is vacated (when the joey is around eight months old).

THREATS: Dingoes, feral cats, Wedge-tailed Eagles and foxes in the southern extent of its range.

Above: Regular grooming eliminates burrs and tangles.

DIET: Grass and some roots

LENGTH: HB 59.3–85 cm; T 58.7–84 cm
WEIGHT: 9–27 kg
STATUS: Secure

Above, clockwise from top left: Taking a pentapedal walk; Males are larger and more muscular than females; This doe's joey will soon be evicted; The thigh stripe is a distinctive feature of the Agile Wallaby; The limbs and tail sometimes appear grey.

Black-striped Wallaby *Macropus dorsalis*

The Black-striped Wallaby seeks safety in numbers. By day, this sociable species rests in permanent camps, where up to 20 individuals congregate. If disturbed, the entire mob will flee, keeping together and moving off in single file, rather than scattering. Most of their movement follows well-established routes and runways through dense vegetation from camps to feeding areas.

FEATURES: This species takes its name from the very prominent black stripe that runs from the forehead down the spine to the rump. The rest of the body is brown with a rufous tinge, often darker brown or grey on the flanks. The digits are black as is the nose and tips of the ears. The Black-striped Wallaby's tail is very sparsely furred with a scaly appearance.

DIET & HABITAT: Prefers the shelter of shrubby undergrowth in acacia woodlands, brigalow scrub, dry forest, lantana thickets and sometimes the margins of rainforest. The Black-striped Wallaby is cautious and rarely ventures out into open territory unless there is dense cover nearby. It feeds on grass and leaves and sometimes on crops (in areas of its range that have been taken over by agricultural land).

BEHAVIOUR: Although most individuals seek the safety of the mob, elderly males are often solitary.

BREEDING: A single joey is born after a gestation period of 33–35 days and stays in the pouch for about 30 weeks. After giving birth, the female quickly comes into oestrus and mates again. Black-striped Wallabies can employ embryonic diapause if environmental conditions are poor. Females are mature at about fourteen months — well before the males, which are not mature until 20 months of age.

THREATS: A lot of land in the Black-striped Wallaby's range has been converted to pastoral property, although the species remains common. Dingoes and Wedge-tailed Eagles are native predators.

DIET: Grass, leaves and crops (where available)

LENGTH: HB 53–82 cm; T 54–83 cm
WEIGHT: 6–20 kg
STATUS: Endangered (NSW); Secure elsewhere

Opposite: A young Black-striped Wallaby.
Above, top to bottom: The dark muzzle and black stripe over the forehead and down the spine characterise this species; Unless there is cover nearby, this species rarely strays into the open; The rufous tinge is obvious on this Black-striped Wallaby.

Tammar Wallaby *Macropus eugenii*

The Tammar Wallaby is one of the smallest wallaby species and mostly inhabits islands off the southern Western Australian and South Australian coast, although there are also some small mainland populations. A now-extinct population on Flinders Island in South Australia had shorter, sleeker fur and was considerably different from the present-day type.

FEATURES: This stout little wallaby has a particularly shaggy, grey-brown coat with a reddish tinge on the limbs and flanks. The underparts are grey-silver and similarly furry. Relatively small, rounded ears and an indistinct white stripe along the jaw are also common. Sometimes a darker dorsal stripe down the centre of the forehead is observed. The thick, comparatively short tail is uniformly grey-brown.

DIET & HABITAT: Prefers coastal heath and scrub with plenty of sheltering ground cover but will also frequent dry sclerophyll forests that have a dense understorey of grasses. Mainland populations are now restricted to Dryandra State Forest and the Tutanning and Perup Nature Reserves, although island populations are more stable. Tammar Wallabies mostly graze on grassy clearings close to dense cover.

BEHAVIOUR: This species is mostly solitary and strictly nocturnal. Small groups may gather in good feeding grounds after dark. Tammar Wallabies use runways through dense cover to reach feeding grounds. Incredibly, Tammar Wallabies have been observed drinking seawater in some of their more arid island habitats — scientific tests even demonstrated that females nourished only on dry grass and saltwater could still suckle young and maintain their body weight.

BREEDING: This species breeds only seasonally. Most offspring are produced from late January to March. Females mate again shortly after giving birth but practise embryonic diapause until after 22 December. Day length and the summer solstice influence this cycle.

THREATS: Mortality of juveniles is as high as 40%, mostly from starvation or cold weather. Feral cats and scrub clearing may have also played a part in their mainland decline.

Left: Female Tammar Wallabies are mature at nine months but males don't mature until two years of age.

DIET: Grass

LENGTH: HB 52–68 cm; T 33–45 cm
WEIGHT: 4–10 kg
STATUS: Near Threatened (IUCN Red List); Endangered (SA)

Above: The Tammar Wallaby's dense coat protects against cold.

Western Brush Wallaby *Macropus irma*

The Western Brush Wallaby, sometimes also called the Black-gloved Wallaby, is one of the larger wallaby species. In some ways it lives a similar lifestyle to the larger kangaroo species — it favours open woodland and floodplains with low grasses and sheltering thickets of scrub. It is primarily a grazer.

FEATURES: Males and females do not differ as much in size as most other wallaby species. The upper body and chest are grey and the underparts a yellowish buff. A white stripe extends from each side of the nose to the black-tipped ears. Forepaws, hindfeet and tip of the crested tail are also black.

DIET & HABITAT: Although sometimes found in mallee habitat and coastal heathland, this species prefers open forest or woodland where seasonal rain brings fresh green pick. The Western Brush Wallaby seems not to require water when environmental conditions are favourable.

BEHAVIOUR: Western Brush Wallabies are solitary and diurnal. They are most active in the early morning and late afternoon and spend the day resting underneath bushes or in shrubby thickets.

BREEDING: Further research into their reproductive behaviour is required, but births appear to peak in April and May, suggesting they may not breed year-round. They probably also practise embryonic diapause. A single offspring is born and stays in the pouch for 6–7 months.

THREATS: Predation by foxes appears to significantly affect this species. Surveys conducted in the late 1990s showed a decline in Western Brush Wallaby numbers, despite larger kangaroo species remaining constant. After fox-baiting, populations appeared to stabilise. Clearing of habitat could also impact on this species. The Western Brush Wallaby became fully protected in 1951 and is now listed as Near Threatened on the IUCN Red List.

Above: Obvious black "gloves" led to this species' alternative common name, Black-gloved Wallaby.

DIET: Grass

LENGTH: HB 90–120 cm; T 54–97 cm
WEIGHT: 7–9 kg
STATUS: Near Threatened (IUCN Red List)

Above, top to bottom: Although solitary, the Western Brush Wallaby can sometimes be seen feeding in pairs; This species has a characteristic gait, head angled low, rump high and tail held low; The classic pentapedal walk.

Parma Wallaby *Macropus parma*

By the mid-1960s, the small Parma Wallaby was presumed extinct in Australia. Thankfully, across Bass Strait, introduced wallaby species (believed at first to be solely the Tammar Wallaby) were wreaking havoc on New Zealand pine plantations. An attempt to control the pests revealed two species, the Tammar and the Parma Wallaby. The Parma Wallaby was gratefully repatriated to Australia, and it was only later, in 1967, that surviving local individuals were discovered near Gosford in New South Wales.

FEATURES: This is the smallest of Australia's wallaby species. Its grizzled grey-brown fur sometimes has a slight rufous tinge, particularly on the flanks. The underparts are grey, although the chin, throat and upper chest are white. A white stripe runs along the cheeks and terminates under the eyes. A dark dorsal stripe runs along the spine and ends in the middle of the back. The tail frequently ends in a white tip. It can easily be confused with pademelons, although they carry their tails stiffly straight out when hopping, whereas the Parma Wallaby's tail forms a "U" shape.

DIET & HABITAT: The Parma Wallaby makes its home in the dense, scrubby understoreys of wet or dry sclerophyll forests (and, infrequently, rainforests) close to grassy areas, where it travels at night to feed on grass, herbs and ferns.

BEHAVIOUR: These nocturnal and usually solitary macropods are sometimes observed feeding in pairs.

BREEDING: Females are sexually mature at one year (2–3 years in the Kawau Island population) and males at 20–24 months. From February to June, females give birth to a single offspring following a gestation period of approximately 35 days. The joey will exit the pouch permanently at about 30 weeks of age but continue to suckle for a further 10–14 weeks.

THREATS: Dingoes, feral cats, Wedge-tailed Eagles and habitat destruction threaten this species.

Above: The timid Parma Wallaby is best observed with binoculars or though a zoom lens. **Opposite, clockwise from top:** Its size often sees it confused with pademelons; The Parma Wallaby has a uniform rufous colour, which is not confined to the neck or legs; Parma Wallabies may feed in pairs.

DIET: Grass, herbs and ferns

LENGTH: HB 44.7–52.8 cm; T 40.5–54.4 cm
WEIGHT: 3.2–5.9 kg
STATUS: Near Threatened (IUCN Red List); Vulnerable (NSW)

Whiptail Wallaby *Macropus parryi*

Also sometimes referred to as the Pretty-face Wallaby, this is one of the most attractive wallaby species. It is also a gregarious and sociable wallaby species, living in large mobs where strict rules of social hierarchy are observed.

FEATURES: The most defining feature of this species is its relatively slender face, which is delineated by an obvious white cheek stripe. A characteristic white band also runs across each ear just below the black ear tip and a white hip stripe extends from the flank to the pale white-grey tail, which terminates in a darker tip. The body colour is a light grey in winter, becoming slightly darker in summer.

DIET & HABITAT: The Whiptail Wallaby subsists on herbaceous plants, grass and ferns in undulating open forest along Australia's eastern coastline from northern New South Wales to similar habitats as far north as Cooktown in Queensland. It is at its most abundant in the southern areas of its range. It rarely needs to drink, unless in times of drought; usually dew and its diet provide it with sufficient water.

BEHAVIOUR: One of the more diurnal macropod species, the Whiptail Wallaby usually feeds at dawn and in the early morning, and also in the late afternoon and evening. Up to 50 Whiptail Wallabies form a mob, within which numerous subgroups form. Each subgroup is controlled by a dominant male and each group's home range overlaps with the home ranges of other groups. Submission is shown by a soft cough, and defensiveness by a hiss and a growl. A soft cluck is made by courting males.

BREEDING: Mating occurs on a single day in the female's 41–44-day oestrous cycle, during which the dominant buck and other males follow her around. Gestation takes 34–38 days.

THREATS: Clearing of forest, raptors, Dingoes and feral cats.

Above: Juveniles reach independence at fifteen months of age. **Opposite, clockwise from top left:** Pouch life lasts about 37 weeks; Dominant bucks must be strong enough to aggressively defend does from younger males; After leaving the pouch, a joey still suckles until it is about fifteen months old.

DIET: Grass, herbs and ferns

LENGTH: HB 75.5–92.5 cm; T 72.8–104.5 cm
WEIGHT: 7–26 kg
STATUS: Secure

Red-necked Wallaby *Macropus rufogriseus*

*The Red-necked Wallaby is a common inhabitant of forest, woodland and coastal heath along Australia's south-east coast and in Tasmania, where it becomes a subspecies commonly referred to as Bennett's Wallaby (*Macropus rufogriseus rufogriseus*).*

FEATURES: Slightly speckled grey-brown fur gives this species a grizzled appearance. On the neck and upper forearms, the fur takes on a reddish-brown tinge. A white cheek stripe runs from the muzzle to below the eyes. The muzzle, paws and outside edges of the ears are black. A black stripe runs down the centre of the forehead but does not continue onto the back. The chin, chest and underparts are pale grey to white. The Bennett's Wallaby subspecies has slightly darker, fluffier fur. The reddish tinge on the neck is less pronounced.

DIET & HABITAT: Primarily a grazer, the Red-necked Wallaby survives mostly on herbs and grasses, although in some areas it will reach pest proportions feeding on crops of pastoral properties. It is most commonly seen in tall eucalypt forests that provide moderate shrubby cover, as well as in tall coastal heath communities.

BEHAVIOUR: Red-necked Wallabies can be considered a solitary species, despite groups of 30 or more feeding on particularly grassy areas. They rest alone by day, although a female and her immature juvenile will rest together.

BREEDING: Unusually for a subspecies, Bennett's Wallaby exhibits different breeding behaviour to the Red-necked Wallaby. Red-necked females give birth all year round, peaking in summer. Bennett's Wallaby females give birth only from late January to July (and females use embryonic diapause outside of these times). Oestrous and gestation are the same for both types, at 33 days and 30 days respectively.

THREATS: Feral cats, and Dingoes are mainland predators. While this species is abundant and secure, habitat destruction is a concern.

Above: Twins are uncommon, but the doe's four nipples enable some to survive. **Opposite, clockwise from top:** Females with pouch young mate and keep a quiescent blastocyst in the uterus; Joeys that have left the pouch may continue to suckle until they are seventeen months old; An alert pose.

DIET: Grass, herbs and some seeds

LENGTH: HB 65.9–92.3 cm; T 62.3–86.2 cm
WEIGHT: 11–26.8 kg
STATUS: Rare (SA); Secure elsewhere

Antilopine Wallaroo *Macropus antilopinus*

The Antilopine Wallaroo is probably frequently confused with the Grey or Red Kangaroo, as it shares similarities with both of these species. It is more slender and kangaroo-like than most other wallaroo species, which are generally shorter and stouter. It also differs from other wallaroo species in that it is gregarious and forms mobs of eight to 30 individuals.

FEATURES: One feature that sets this species apart from the Grey and Red Kangaroo species is its white-tipped ears. The Antilopine Wallaroo's fur is otherwise a reddish-tan that is often greyer on the shoulders. The underparts are a pale grey to white, as are the limbs and tail (although the paws are black). An indistinct white stripe or patch may be seen on the cheek, but is not consistent. Females are often more uniform in colour and may be all grey or all tan.

DIET & HABITAT: Prefers eucalypt woodlands with a grassy understorey on flat plains or gently sloping hills, but is also seen in some rocky terrain that is usually dominated by the Common Wallaroo. The Antilopine Wallaroo feeds mostly, possibly entirely, on grasses.

BEHAVIOUR: Most mobs contain about eight individuals and it is likely that larger groups form only to take advantage of food or when threatened by predators such as the Dingo. They are nocturnal and conserve energy in the dry season by resting under shady bushes by day, often close to waterholes or billabongs. In the wet season, the Antilopine Wallaroo is more diurnal and may be seen by day.

BREEDING: The gestation period is approximately 34 days and offspring are produced throughout the year. Unusually, females of this species do not appear to mate again immediately after giving birth. Joeys remain in the pouch for around nine months.

THREATS: Dingoes, Wedge-tailed Eagles and White-bellied Sea-Eagles in some parts of their range.

Above: The characteristic short, blunt head and white-tipped ears. **Opposite, clockwise from top left:** Individuals keep watch to alert the mob when danger threatens; A juvenile close to independence; Antilopine Wallaroos rest during the heat of the day.

DIET: Almost entirely grass

LENGTH: HB 77.8–120 cm; T 67.9–89 cm
WEIGHT: 16–49 kg
STATUS: Secure

Black Wallaroo *Macropus bernardus*

Easily identified by its stocky, short body and dark fur, the Black Wallaroo is Australia's smallest and most timid wallaroo species. This species is also unique for being the only Macropus *species to have eighteen chromosomes (compared with twenty for the Red Kangaroo and sixteen for other species in this genus). Its distribution is limited and it exists only in a small area of the Northern Territory's Arnhem Land escarpment and plateau.*

FEATURES: Males are stocky, muscular and darker than females, which are generally grey or grey-brown (while males are dark brown to black). The paws, feet and tip of the tail are dark brown or black on both sexes. The ears are small and rounded.

DIET & HABITAT: Arnhem Land's rocky escarpments and steeply rising plateaus covered with eucalypt woodland are this species' preferred habitats. In this monsoonal climate Black Wallaroos especially favour hummock grassland and woodland with an understorey of grass, shrubs and sometimes spinifex. Occasionally they are associated with small patches of rainforest. When disturbed in open country, they will quickly retreat to their preferred rocky habitat. Diet consists mostly of grass, shrubs and ground plants and this species may be seen foraging on road verges. The Black Wallaroo requires a reliable water supply.

BEHAVIOUR: Black Wallaroos are solitary and are only ever seen in groups of three (usually a female and her juvenile joey and an adult male). They are usually inactive by day and feed in the evening — although on overcast days in the wet season they may be seen during the daytime. Black Wallaroos are also notoriously shy and flee when they are approached.

BREEDING: Reproduction has not been studied in detail, but breeding possibly peaks in late summer.

THREATS: Dingoes, drought and fire may threaten this species.

Above: Males are especially square-jawed with broad faces and muscular chests. **Opposite, clockwise from top left:** Females are paler, but still stockier than grey kangaroos; Some males grow very dark; The thick tail helps balance; Rocky wallaroo habitat in Arnhem Land.

DIET: Grass, shrubs and ground plants

LENGTH: HB 59.5–72.5 cm; T 54.5–64 cm

WEIGHT: 13–22 kg

STATUS: Near Threatened (IUCN Red List)

Common Wallaroo *Macropus robustus*

This solid, long-eared wallaroo species was first recorded in drawings made by Joseph Banks in 1770, while the Endeavour *was being repaired near Cooktown in north Queensland. John Gould later described the species in 1839 from two specimens collected in New South Wales.*

FEATURES: The Common Wallaroo has dark grey-brown fur, often with a rufous tinge, particularly on the flank and back. Females often have blue-grey or brown body fur. Underparts are pale grey to silver, as are the forelimbs and hindlimbs, although the paws and toes are reddish black. The nose is black and the ears are exceptionally large and flared. Often the long, thick tail is tinged red-orange. In the west of its range (where the subspecies *Macropus robustus erubescens* is commonly known as the Euro), the Common Wallaroo's fur may be shorter and more rufous.

DIET & HABITAT: The Common Wallaroo, as the name suggests, is widely distributed, ranging over most of the mainland. Its habitat is extremely varied but mostly includes rocky terrain, hills and escarpments where it can hide and escape hot weather. This species can also survive with very little water (provided it can take shelter from the sun and consume succulent plants), increasing its range to Australia's most arid habitats. Its preferred diet is grass, shrubs and succulent ground plants.

BEHAVIOUR: Common Wallaroos are nocturnal and solitary. If alarmed, the Common Wallaroo will emit a loud hiss as it exhales. Another common vocalisation is a loud "cch-cch" noise.

BREEDING: Offspring are produced year-round, but breeding may cease during drought. Gestation time differs between the Common Wallaroo and the Euro.

THREATS: Domestic livestock may compete with this species. Dingoes and Wedge-tailed Eagles are its main predators.

Above: A juvenile adopts the typical wallaroo stance.

DIET: Grass, ground plants and shrubs

LENGTH: HB 110.7—198.6 cm; T 53.4–90.1 cm
WEIGHT: 6.25–46.5 kg
STATUS: Subspecies are Vulnerable in parts

Above, clockwise from top: Common Wallaroos prefer rocky habitats; The Common Wallaroo has a more upright gait than kangaroos; Although this species can survive with little water, it needs good cover to avoid dehydrating in hot weather.

Western Grey Kangaroo *Macropus fuliginosus*

Australia's two species of grey kangaroo — Eastern Greys and Western Greys — were the subject of much conjecture among early taxonomists. Both species are widespread, so at one time they were believed to be as many as five grey kangaroo species. Science has now confirmed that there are just two species, which are distinguished largely by reproductive biology. Female Western Grey Kangaroos, unlike their Eastern Grey counterparts, do not exhibit embryonic diapause and have shorter oestrous and gestation periods than Eastern Greys.

FEATURES: Superficially, Western Greys appear very similar to Eastern Greys, although the face and upper body are usually darker brown in colour. Generally, Western Greys are also more thickly coated. The underparts are a pale fawn to buff brown. The tips of the hindpaws and forepaws are darker brown to black, as is the tail tip. Males are sometimes colloquially known as "stinkers" for their characteristically pungent body odour.

DIET & HABITAT: Western Greys occupy the southern and western parts of the mainland, with populations spread throughout southern Western Australia, South Australia, western Victoria, north-western New South Wales and south-western Queensland. A subspecies *Macropus fuliginosus fuliginosus* is found on Kangaroo Island in South Australia. Western Grey Kangaroos prefer flat plains and open woodland with a grassy understorey. They feed mostly on grass, broad-leaved ground plants and some shrubs. Agriculture has significantly increased populations in some parts of their range and licences for strictly controlled culling are sometimes granted in order to reduce their numbers.

BEHAVIOUR: Western Greys are gregarious and form large mobs with a dominant buck as the "leader". A crepuscular lifestyle suits these macropods, which are most active in the early morning or late afternoon.

BREEDING: After a gestation period averaging 30.5 days, a single joey is born and stays in the pouch for about 42 weeks — about fourteen weeks less than the pouch life of the Eastern Grey. The female's oestrous cycle is also shorter, averaging 35 days.

THREATS: Dingoes, cats, Wedge-tailed Eagles and habitat destruction.

Left: Joeys are independent at twenty months.

DIET: Grass, ground plants, shrubs and crops (when available)

LENGTH: HB 52.1–122.5 cm; T 42.5–100 cm
WEIGHT: 3–53.5 kg
STATUS: Secure

Above, top to bottom: Darker, fluffier fur distinguishes this species from the Eastern Grey; Like Eastern Greys, the Western Grey forms social mobs.

Eastern Grey Kangaroo *Macropus giganteus*

Eastern Grey Kangaroos inhabit coastal plains along Australia's eastern coast in Queensland, New South Wales, Victoria, Tasmania and the south-eastern corner of South Australia. They are slightly paler than Western Greys, with which they share a small part of their range.

FEATURES: Eastern Greys are slightly less fluffy than Western Greys and are usually a grey-brown to silver grey colour with pale, light grey underparts. The tips of the forepaws and hindfeet are slightly darker brown, as is the tip of the tail. The Eastern Grey's large, oblong ears are often fringed with white hairs in their inner corners.

DIET & HABITAT: Usually, the Eastern Grey inhabits coastal woodland, plains, the undulating landscape of the Great Dividing Range and areas along the eastern coast where rainfall is greater than 250 mm per annum. They have also been recorded in mallee scrub and tea-tree woodlands, and are frequent visitors to pastoral properties. Increased availability of groundwater, by way of dams and water troughs, has benefited this species. Eastern Grey Kangaroos subsist mostly on grasses and forbs.

BEHAVIOUR: Eastern Greys are sociable and form large mobs of up to 100 individuals. Males and females within the mob have separate hierarchies based on age and reproductive standing. This species is also crepuscular, venturing out to feed in large groups in the early morning and evening. Eastern Greys make a number of vocalisations. Males wooing a female in oestrus mimic the soft clucks made by mothers to their young. When alarmed, both males and females utter a guttural coughing noise.

BREEDING: Gestation takes about 36 days and births occur year-round, peaking in summer. Twins are rarely produced, but have been recorded. Usually a single joey is born and vacates the pouch eleven months later.

THREATS: Dingoes, cats, raptors and habitat destruction.

Above: The muzzle is finely furred with hair between the upper lip and nostrils. **Opposite, top to bottom:** Eastern Greys stay alert while resting; Eastern Greys can employ embryonic diapause when they have pouch young; A beachside mob feeds at Murramarang National Park, New South Wales.

DIET: Grasses, forbs and crops

LENGTH: HB 51.2–121.2 cm; T 43–109 cm
WEIGHT: 3.5–66 kg
STATUS: Secure

Red Kangaroo *Macropus rufus*

The Red Kangaroo is the largest marsupial in the world — standing over 1.4 m tall. It is the only kangaroo species truly suited to Australia's harsh, arid habitats, but prefers greener plains and open woodlands with plenty of fresh green herbaceous plants.

FEATURES: Male Red Kangaroos ("boomers") are heavily muscled with a rich, rusty red coat, although some may also be a dark grey colour. Females (often called "blue flyers") are much smaller and are grey-blue to reddish. Colours vary considerably with location and season. A broader snout and white cheek stripe extending to beneath the ears, along with their larger size, distinguishes Red Kangaroos from Eastern Greys and Western Greys.

HABITAT: Grasslands and woodlands in arid or semi-arid areas make up much of this kangaroo's range across Central Australia; however, they also inhabit regions with better rainfall. Dams, bores and inland water troughs have benefited this species considerably — providing water for longer periods in seasonally dry areas. Their preferred diet is fresh green grass, native herbs and the foliage of some shrubs.

BEHAVIOUR: Red Kangaroos are mob animals, forming social groups. In areas where food is plentiful, hundreds of them may aggregate to feed. They are largely crepuscular, feeding at dawn and dusk, and conserving energy by resting in the shade during the day. They regulate their body heat in hot weather by panting and licking the insides of their forearms.

BREEDING: Red Kangaroos are slow developers compared with most other macropods, with both males and females reaching sexual maturity at 2–3 years of age. Females exhibit embryonic diapause and may have three joeys in different stages of growth at any one time — a quiescent blastocyst, a joey in the pouch (drinking from one teat) and a joey at foot that suckles from another teat. Females produce different milk for pouch young and young at heel.

THREATS: Dingoes and Wedge-tailed Eagles are natural predators. Annual quotas are set for shooters.

Above: Wild dogs and Dingoes, which often hunt in packs, are the only natural predators of fully grown Red Kangaroos. Joeys are also vulnerable to pythons and birds of prey.

DIET: Grasses, forbs and crops

LENGTH: HB 74.5–140 cm; T 64.5–100 cm
WEIGHT: 17–85 kg
STATUS: Secure

Above, top to bottom: Artificial water sources, such as dams and bores, have allowed Red Kangaroos to extend their range almost right across Australia; When environmental conditions are good, populations skyrocket and quota-based, licensed hunting is introduced; Powerful, muscular hindlimbs propel the Red Kangaroo with speed and grace. **Inset:** A Red Kangaroo doe and joey.

ACACIA Both the genus and common name for many wattle species.

ARBOREAL Living in trees.

BIPEDAL (GAIT) Two-limbed movement.

BLASTOCYST Early embryo stage in mammal development.

BRIGALOW Open forest dominated by *Acacia harpophylla*, 10–15 m high.

CANOPY Highest level of tree branches and foliage.

CHROMOSOME A thread of genes and DNA found in the nucleus of a cell. Different organisms have different numbers of chromosomes.

CLOACA Posterior chamber of the gut in monotremes and marsupials where the urinary tract and female reproductive system end.

CREPUSCULAR Active at dawn and dusk.

DENTITION Relating to the teeth.

DICOTYLEDONOUS (PLANT) Group of flowering plants where the seed typically features two embryonic leaves.

DIURNAL Active during the day.

EMBRYO Animal in developmental stage between conception and birth.

EMBRYONIC DIAPAUSE State of arrested development in a viable embryo, which may be carried in the uterus for some months.

FERAL Having reverted to a wild state.

FORB A special type of flowering plant with a non-woody stem. Not in the general category of a grass, shrub or tree.

GAIT An animal's method and style of movement.

GESTATION Time between conception and birth.

GREGARIOUS Social, living in groups.

HEATHLAND Vegetation dominated by hard-leaved, small shrubs (usually less than 2 m high) growing in poor, sandy soils.

HERBIVORE Animal that eats plants.

HOME RANGE Area an animal traverses during its normal daily activities.

HYBRID Offspring of two different species.

MALLEE Small, multi-stemmed eucalypts that often dominate semi-arid and arid areas.

MARSUPIUM Pouch, a distinguishing feature of most female marsupials.

MONTANE Mountainous areas below the treeline.

NOCTURNAL Active during the night.

OESTROUS CYCLE A period of sexual receptivity in females, caused by reproductive hormones that stimulate the release of an ova (egg) ready to be fertilised by the male.

PENTAPEDAL (GAIT) Five-limbed movement.

PREHENSILE Able to grip.

QUADRUPEDAL (GAIT) Four-limbed movement.

RIPARIAN The ecological zone between land and a flowing body of water.

SCLEROPHYLL (FOREST) Forest dominated by sclerophyllous (hard-leaved) trees, especially eucalypts.

SCREE Collection of broken rocks and rubble that forms at the foot of cliffs, crags and mountain ranges.

SUCCULENT A fleshy or juicy plant.

SUPERNUMERARY More than the usual number of body parts.

TAXONOMY The scientific practice of classifying life.

UNDERSTOREY The shrubs, saplings and plants that grow under the forest canopy.

WOODLAND Area sparsely covered by trees.

A
Aepyprymnus rufescens 23

B
Bettong 4, 12, 18–23
  Brush-tailed 21, 22
  Burrowing 18, 19
  Northern 21
  Rufous 21, 23
  Southern 20, 23
  Tasmanian (see Bettong, Southern)
Bettongia gaimardi 20
Bettongia lesueur 18
Bettongia penicillata 22
Bettongia tropica 21

C
Caloprymnus campestris 12

D
Dendrolagus bennettianus 46
Dendrolagus lumholtzi 47

E
Euro 100

H
Hare-wallaby 26–31
  Banded 31
  Eastern 26
  Rufous 26, 27, 30
  Spectacled 28
Hypsiprymnodon moschatus 24

K
Kangaroo 4, 80, 81, 102–107,112
  Eastern Grey 6, 8, 9, 80, 81, 103–105, 109
  Red 5, 6, 8, 9, 11, 80, 96, 98, 106, 107
  Western Grey 6, 81, 102, 103

L
Lagorchestes conspicillatus 28
Lagorchestes hirsutus 30
Lagostrophus fasciatus 31

M
Macropod 4–11
Macropus agilis 82
Macropus antilopinus 96
Macropus bernardus 98
Macropus dorsalis 84
Macropus eugenii 86
Macropus fuliginosus 102
Macropus giganteus 104
Macropus irma 88
Macropus parma 90
Macropus parryi 7, 92
Macropus robustus 7, 100
Macropus robustus erubescens 100
Macropus rufogriseus 94
Macropus rufogriseus rufogriseus 94
Macropus rufus 106
Monjon 50, 52, 53

N
Nabarlek 50, 52, 54, 55

O
Onychogalea fraenata 76
Onychogalea unguifera 78

P
Pademelon 32–39
  Red-legged 36, 37
  Red-necked 36, 38, 39
  Tasmanian 34
  Red-legged 36, 37
Petrogale assimilis 60
Petrogale brachyotis 50
Petrogale burbidgei 52
Petrogale coenensis 63
Petrogale concinna 54
Petrogale concinna canescens 54
Petrogale concinna concinna 54
Petrogale concinna monastria 54
Petrogale godmani 58
Petrogale herberti 59
Petrogale inornata 61
Petrogale lateralis 64
Petrogale lateralis hacketti 64
Petrogale lateralis lateralis 64
Petrogale lateralis pearsoni 64
Petrogale lateralis purpureicollis 72
Petrogale mareeba 56
Petrogale penicillata 66
Petrogale persephone 70
Petrogale purpureicollis 72
Petrogale rothschildi 73
Petrogale sharmani 62
Petrogale xanthopus 68
Potoroo 4, 12–17
  Broad-faced 12
  Gilbert's 17
  Long-footed 14, 16
  Long-nosed 12, 14, 15
Potorous gilbertii 17
Potorous longipes 16
Potorous platyops 12, 13
Potorous tridactylus 12, 14

Q
Quokka 40, 41

R
Rat-kangaroo
  Desert 12
  Musky 4, 8, 12, 24, 25
Rock-wallaby 48, 50, 56, 58–62, 66–68
  Allied 6, 58, 60, 61, 62
  Black-footed 6, 64, 65, 72
  Brush-tailed 59, 66
  Cape York 58, 63
  Godman's 56, 58, 63
  Herbert's 59, 66
  Little (see Nabarlek)
  Mareeba 6, 48, 56, 57, 58, 61, 62
  Mt Claro (see Rock-wallaby, Sharman's)
  Proserpine 70
  Purple-necked 72
  Rothschild's 73
  Sharman's 61, 62
  Short-eared 50, 51
  Unadorned 59, 60, 61
  Yellow-footed 6, 48, 49, 68, 69

S
Setonix brachyurus 41

T

Thylogale billardierii 34
Thylogale stigmatica 36
Thylogale thetis 38
Tree-kangaroo 44–47
  Bennett's 44, 46, 47
  Lumholtz's 44, 46, 47

W

Wallabia bicolor 43
Wallaby 4, 5, 6, 7, 8, 80–95
  Agile 9, 82, 83
  Bennett's 94
  Black-striped 84, 85
  Black-gloved (*see* Wallaby, Western Brush)
  Nailtail 74–79
  Nailtail, Bridled 74, 76, 77, 78
  Nailtail, Crescent 74
  Nailtail, Northern 74, 75, 78
  Parma 80, 90, 91
  Pretty-face (*see* Wallaby, Whiptail)
  Red-necked 6, 80, 94, 95, 111
  Swamp 6, 42, 43
  Tammar 11, 86, 87, 90
  Western Brush 6, 9, 88, 89
  Whiptail 6, 7, 9, 10, 92, 93
Wallaroo 81, 96–101
  Antilopine 6, 96, 97
  Black 98, 99
  Common 7, 80, 96, 100, 101

Links & Further Reading

Books

Currey, K. *Fact File: Mammals*, Steve Parish Publishing, Brisbane, 2006

Curtis, L. K. *Green Guide: Kangaroos & Wallabies of Australia*, New Holland, Sydney, 2006

Egerton, L. (Ed), *Encyclopedia of Australia Wildlife*, Reader's Digest, Sydney, 2005

Grigg, G. Jarman, P & Hume, I (Eds.) (*Kangaroos, Wallabies and Rat-kangaroos*, 3 vols, Surrey Beatty and Sons, Sydney, 1989

Jones, C. & Parish, S. *Field Guide to Australian Mammals*, Steve Parish Publishing, Brisbane, 2004

Lindsey, T. *Green Guide: Mammals of Australia*, New Holland, Sydney, 1998

Menkhorst, P. & Knight, F. *Field Guide to the Mammals of Australia*, Oxford University Press, Melbourne. 2001

Strahan, Ronald (Ed.) *The Mammals of Australia*, Reed New Holland, Sydney, 2002

Watts, D. *Kangaroos & Wallabies of Australia*, New Holland, Sydney, 1999

Websites

Action Plan for Australian Marsupials and Monotremes **www.environment.gov.au/biodiversity/threatened/publications/action/marsupials/**

Australian Government: Department of the Environment and Water Resources **www.environment.gov.au/biodiversity/trade-use/wild-harvest/kangaroo/**

Australian Museum **www.amonline.net.au**

IUCN Red List of Threatened Species **www.iucnredlist.org/**

Marsupial Society of Australia **www.marsupialsociety.org/**

News in Science **www.abc.net.au/science.news**

The Wilderness Society **www.wilderness.org.au/**

Australian Wildlife Conservancy **www.australianwildlife.org**

Acknowledgements

The publisher wishes to thank the staff of Healesville, Scotia, Yookamurra, Rainforest Habitat and Northern Territory Sanctuaries for providing access to rare and endangered macropods. Also thanks to Dr Les Hall & Ian Morris for checking the facts contained in this book.

Published by Steve Parish Publishing Pty Ltd
PO Box 1058, Archerfield, Qld 4108 Australia
www.steveparish.com.au
© Steve Parish Publishing

All rights reserved. No part of this publication may be reproduced, stored in a retrieval system, or transmitted in any form or by any means, electronic, mechanical, photocopying, recording or otherwise, without the prior permission in writing of the publisher.

ISBN 978174193323 9

First published 2008

Principal photographer: Steve Parish

Photographic assistance: Greg Harm, SPP: p. 74 (bottom)

Additional photography: Martin Harvey/ANTPhoto.com: p. 46; Nicholas Birks/Auscape: p. 107 (top); Jean-Paul Ferraro/Auscape: p. 11 (bottom left); Wayne Lawler/Auscape: p. 72; D Parer & E Parer-Cook/Auscape: p. 11 (top left & right, and bottom right); Becca Saunders/Auscape: p. 64; Stanley Breeden: p. 25 (top); Michael Cermak: pp. 42 (bottom left), 47 & 57 (top); Jiri Lochman/Lochman Transparencies: pp. 17, 52, 53 (top), 54 & 73; Marie Lochman/Lochman Transparencies: p. 31; Dave Watts/Lochman Transparencies: p. 24; Ian Morris: pp. 50, 51 (top), 53 (bottom), 55 (top) & 79 (bottom right); Bruce Cowell/Queensland Museum: p. 21; Gary Cranitch/Queensland Museum: p. 38 & 85 (top); Jeff Wright/Queensland Museum: p. 94; Robert Close/University of Western Sydney: pp. 59 & 62; Martin Willis: p. 51 (bottom)

Illustrations: John Gould: pp. 12, 26 & 74

Text: Karin Cox
Design: Gill Stack, SPP
Editing: Ted Lewis & Michele Perry, SPP; Kerry Davies
Production: Tina Brewster, SPP

Prepress by Colour Chiefs Digital Imaging, Brisbane, Australia
Printed in Singapore by Imago

Produced in Australia at the Steve Parish Publishing Studios